Game Theory and International Relations

Game Theory and International Relations

Preferences, Information and Empirical Evidence

Edited by

Pierre Allan

Professor of Political Science
University of Geneva, Switzerland

and

Christian Schmidt

Professor of Economics
University of Paris-Dauphine, France

Edward Elgar

Published by
Edward Elgar Publishing Limited
Gower House
Croft Road
Aldershot
Hants GU11 3HR
England

Edward Elgar Publishing Company
Old Post Road
Brookfield
Vermont 05036
USA

British Library Cataloguing in Publication Data
Game Theory and International Relations:
Preferences, Information and Empirical
Evidence
I. Allan, Pierre II. Schmidt, Christian
327.101

Library of Congress Cataloguing in Publication Data
Game theory and international relations: preferences, information, and empirical evidence/ [edited] by Pierre Allan and Christian Schmidt.
p. cm.
1. International relations. 2. Game theory. I. Allan, Pierre.
II. Schmidt, Christian.
JX1391.G28 1994
327—dc20 93–50871
CIP

ISBN 978 1 85278 925 1

Printed and bound by CPI Group (UK) Ltd, Croydon, CR0 4YY

Contents

Notes on the editors

Pierre Allan is Professor of Political Science at the University of Geneva, Head of its Political Science Department, and an Executive Committee Member of the Swiss and the International Political Science Associations. He was Visiting Professor at Economics University, Prague in 1993 and at the University of California at Berkeley in 1993–4. His recent published works include *The End of the Cold War: Evaluating Theories of International Relations* (co-editor) and *Horloges sociales, confrontations et invention* (co-author, forthcoming).

Christian Schmidt is Professor of Economics at the University of Paris-Dauphine, Director of the LESOD (Laboratoire d'Economie et de Sociologie des Organisations de Défense) and President of the International Defense Economics Association. He is a member of the editorial board of the *Journal of Conflict Resolution* and *Defence Economics*. He has published various books on the relationship between economics, defence and security matters; among them are *The Economics of Military Expenditures* (editor) *Peace, Defense and Economic Analysis* (co-editor) and, more recently, *Penser la guerre, penser l'économie*.

Notes on the contributors

Vinod K. Aggarwal is a professor in the Department of Political Science, Affiliated Professor at the Haas School of Business, and Chairman of the Political Economy of Industrial Societies Program at the University of California at Berkeley. Among other works on the politics of trade and finance, his books include *Liberal Protectionism* and *International Debt Threat* and the forthcoming books *Horloges sociales, confrontations et invention* (co-author) and *Debt Games: Strategic Interaction in International Debt Rescheduling*.

Lars-Erik Cederman holds an MSc in Engineering Physics from Uppsala University, Sweden, and an MA from the Graduate Institute of International Studies, Geneva. He is currently a doctoral candidate in political science at the University of Michigan, Ann Arbor, where he is a MacArthur Fellow in International Peace and Security.

Cédric Dupont is a PhD candidate in international relations at the Graduate Institute of International Studies in Geneva. He currently teaches international relations at the University of Geneva and has held a research appointment at the University of California at Berkeley. He has published articles in global modelling, international negotiations and political economy.

Urs Luterbacher is Professor of Political Science at the Graduate Institute of International Studies in Geneva. He earned his PhD from the University of Geneva in 1972. He is the author/co-author of *Dynamic Models of International Conflict*, *Co-operative Models in International Relations Research* and *Pathways of Understanding: The Interactions of Humanity and Global Environmental Change* and of numerous articles on international political and environmental problems.

Michael Nicholson was educated at Trinity College, Cambridge, 1953–8 (MA and PhD). He is the author of various books and papers on economics and international relations, most recently *Formal Theories in International Relations* and *Rationality and the Analysis of International Conflict*. For many years he was Director of the Richardson Institute for Conflict and Peace Research and more recently Professor of International Relations at the University of Kent. He is now Professor of International Relations at the University of Sussex.

Gerald Schneider has studied at the universities of Aarhus, Michigan, and Zurich where he obtained his PhD in political science. Since 1992 he has been a lecturer at the Graduate Institute of International Studies in Geneva, Switzerland. His publications include three books (*How Far Do Governments Look Ahead?*, *Die Informationsbeschaffung des Gesetzgebers* and *Time, Planning, and Policy-making*) and ten scholarly articles on international relations and comparative politics. He is currently writing a book-length study on the theory of political integration.

Jean-Paul Theler graduated in June 1991 as a Master of Science in Econometrics and Mathematical Economics from the London School of Economics and Political Sciences. In 1991 he worked as a research assistant at the Créa Institute of Applied Macroeconomics at the University of Lausanne. Since then, he has been a teaching assistant for graduate students in economics at the University of Lausanne and is presently working on his PhD thesis in economics.

1. Introduction

Christian Schmidt and Pierre Allan

William Riker states that the first article ever devoted to game theory in a journal of political science was a paper written by Lloyd Shapley and Martin Shubik, 'A method of evaluating the distribution of power in a committee system' in the *American Political Science Review* in 1954.[1] This publication, only ten years after the first edition of John von Neumann and Oskar Morgenstern's *Theory of Games and Economic Behavior* (1944) suggests that game theory was applied almost simultaneously to political science and economics. Let us note, however, that the two authors of the paper were not trained political scientists, but economists. Curiously enough, in retrospect, the topic chosen does not belong to international affairs but to internal politics. Von Neumann and Morgenstern only developed the theory of co-operative games which, at least at first glance, seems easier to apply to internal than to external politics.

Matters have changed since John Nash's seminal contribution to non-co-operative games (1950, 1951). Non-co-operative games provide a relevant starting point for studying many international situations where interaction between countries does not imply co-operation or even co-ordination, as in the case of warfare, the arms race and some negotiations. Nash's findings were used in international relations long before they were extensively applied in economics. None the less, some professional economists (Martin Shubik, Thomas Schelling, Michael Intriligator and Dagsbert Brito) reached various interesting results by using game theory.

The time has now come to assess how successful game theory can be when applied to international relations. There is no question today that the class of non-co-operative games provides a simple and elegant framework for depicting situations where there exists a combination of conflict and co-operation among countries. By using the concept of Nash equilibrium and other related concepts, one can thoroughly explore a number of logical properties pertaining to these types of situation. Unfortunately, oversimplification represents a common risk and causes much scepticism. This is due in part to the main current of present game theory. For this reason, more and more political scientists are now exploring new paths in game theory, such as sequential and differential games. Because it is generally difficult to measure concepts in international politics the way one measures prices and quantities in economics, one may wonder how relevant cardinal indices are. To escape this traditional criticism of numerical values chosen for pay-offs, it makes sense to use qualitative games, which are often more convenient in modelling international affairs.

There are even more fundamental limitations in game theory itself, many of which relate to the treatment of preferences. Let us single out some of the theory's most frequent assumptions. In game theory, (a) each player chooses his or her strategy by reference to a well-defined preference order in the set of outcomes, and (b) players' preferences are independent. Generally speaking, game theorists consider players' preference orders as a given element and focus their attention on the best way to satisfy them by choosing their strategy according to assumption (a). Unfortunately, the preference order of a country is rather more complicated to identify. It is not a given element because it derives most often from a complex compromise between different preferences linked to internal, as well as external, constraints. Applying game theory to international topics therefore requires further investigations that go beyond the traditional treatment of preference in game theory. Furthermore, countries may not be neutral *vis-à-vis* the others. In that case, countries' preferences ought to be defined as interdependent, and this contradicts assumption (b). The conventional definition of players' preferences must then be revised so as to integrate the hostility or benevolence of each country *vis-à-vis* the others in an international game.

A final difficulty relates to the methodological issues that arise from applying game theory to international relations. According to the data available, the game theorist is usually able to shape the situation he observes into different and alternative types of models. Let us take the Cuban missile crisis in 1962 and, more recently the Gulf War, as two examples. In both cases the variables have been arranged so that the events may be modelled in different ways.[2] The first problem then raised by game theorists is that of selecting the appropriate criteria to choose the better game to explain a situation. Analysing past cases is helpful in providing an initial solution to this problem: one compares the game that was actually played with the set of potential games consistent with the initial data. Yet another possibility is to adopt the viewpoint of the players themselves in the light of their own available data. Players may not have sufficient information to know clearly the game they are playing, but the extent of their knowledge is sufficient to identify a set of possible games. Then comes the problem of investigating the relevant learning processes during which one gropes for the right game among those belonging to the set.

The papers collected in this volume attempt to solve some of these problems. They are grouped in three parts according to the main topic they deal with. Some contributions are more theoretical than others. All the papers, however, seek to balance theoretical considerations with empirical illustrations or in-depth case studies as they try to improve the explanatory possibilities of game theory in international relations. Several chapters raise the issue of the empirical evidence and attempt – through careful analyses of case studies – to show the relevance of specific game theoretical constructs.

Part I explores how actors' preferences are formed and, more precisely, how their different facets can be aggregated in a national or collective preference order. Both Aggarwal and Allan's joint paper and Cederman's paper underline that collective preferences must be viewed as the consequence of some collective rules applied to individual orders. For Aggarwal and Allan, collective preferences are the result of the political actors' constraints related to: (a) their overall position, (b) their specific power, (c) the strength of their domestic coalition. For Cederman, a national interest associated with each country emerges from a process of aggregation of individual orders, corresponding to the organizations involved in international politics. Aggarwal and Allan apply their theory of 'constrained preferences' to the study of the Polish debt negotiations with international banks since 1981 and test its relevance for predicting the negotiated outcomes throughout the 1980s and early 1990s. Cederman's contribution sets forth a logical protocol to generate a well-defined 'national interest'. He envisages this aggregate procedure as a general starting point for studying rational strategic decisions in the arena of international affairs. Both papers use a qualitative approach for framing international situations in non-co-operative games.

Part II focuses on broadening the definition of the preferences of countries. Nicholson's and Schmidt's contributions propose moves from independent to interdependent preferences. Nicholson's paper assumes that in international life, players are concerned by the happiness or unhappiness of the other players that results from the outcome of the game. Nicholson then introduces the concepts of 'benevolence' and 'malice' to measure the pleasure or pain derived by each player at the other player's expense. He also gives an original definition of 'dominance' which he calls an inherent desire to win, independently of any other consideration. He mainly analyses the transformation of an initial non-co-operative game into another when values such as 'benevolence', 'malice', or 'dominant' coefficients are introduced. Schmidt's paper is devoted to investigating crisis phenomena in the process of international decision-making. He also starts by assuming that the preferences of national decision-makers are interdependent, and demonstrates that the multiplicity of non-co-operative games is consistent with the principle of rationality. He draws up a typology of crises in international decision-making systems. Crises due to 'instability', 'undecidability' and 'perverse rationality' are identified through the existence (or lack) of pure Nash equilibria strategies and their related properties.

Part III is mainly concerned with dynamics and the role of actors' information in evolutionary international politics. Schneider's and Dupont's contributions use formal sequential multi-stage games, whereas Luterbacher and Theler's joint contribution opts for differential games. They all consider international negotiations: Schneider's paper investigates the European integration process according to its various phases, namely pre-negotiation, integration negotiation with incomplete information, and ratification process with uncertainty. Starting with a Stackelberg Cournot sequential game, Schneider expands his analyses to two other non-co-operative games based on the steps of the negotiation. He highlights the importance of the intricate relationship between domestic constituents and national governments involved in the negotiation to explain the different pace at which nations enter the European integration process. Dupont's paper describes international negotiation through sequential bargaining models with incomplete information. He shows that the result obtained is different according to (a) the structure of the information, (b) the 'hard' or 'soft' behaviour adopted by the players. His models are accompanied by two case studies: the negotiation over Swiss accession to the League of Nations over the period 1919–20, and the negotiation towards Swiss accession to the European Economic Area of the European Community in the early 1990s. The results are then discussed from a methodological point of view. Luterbacher and Theler's joint paper deals with negotiations about migrations between northern and southern countries. The basic model proposed by the authors puts economic information, such as social welfare functions and standard Cobb–Douglas production functions, into a dynamic two-person game, where each country maximizes an integral of its welfare function. Because of this framework, the results demonstrated are independent of a particular discount rate. Furthermore, it provides an interesting bridge between game theory and optimal control, which in turn allows for the study of conditions that drive both countries to their maximum value. Finally, Luterbacher and Theler demonstrate that the dynamic bargaining situation between North and South takes the form of a prisoner's dilemma.

Contributions to this volume have been organized according to two criteria. First, that there exists a logical development from preference formation to rational choice and from rational choice to dynamic games. Secondly, within each part, the papers are in order of increasing technical complexity. Thus, the reader who is not familiar with the subtleties of game theory may start the first chapter of each part and then read the others for further interest.

NOTES

1. W.H. Riker, 'The entry of game theory into political science', in E.R. Weintraub (ed.), Toward a History of Game Theory, Raleigh, N.C.: Duke University Press, 1992, pp. 207–30. For a more analytical view, see P. Ordeshook, *Game Theory as Political Theory*, Cambridge: Cambridge University Press, 1986.
2. As for modelling the Cuban Missile Crisis, the relevance of the game of chicken, prisoner's dilemma or other games has been extensively discussed by S. Brams, *Superpower Games*, New Haven, Conn.: Yale University Press, 1985, pp. 48–85, and *Negotiation Games*, London: Routledge, 1990, pp. 101–37. Concerning alternative games for modelling the Gulf War, see

C. Schmidt, *Penser la guerre, penser l'économie*, Paris: Odile Jacob, 1991, pp. 226–41, and R. Rudnianski, 'Dissuasion, perception et rationalité, la décision irakienne d'envahir le Koweït: une approche par la théorie des jeux', in *L'Aide à la décision dans la crise internationale*, Paris: Presses de la Fondation pour l'Etude de la Défense Nationale, 1992.

REFERENCES

Nash, John F. (1950), 'Equilibrium Points in *n*-person Games', *Proceedings of the National Academy of Sciences USA*, **36**, pp. 48–9.

Nash, John F. (1951), 'Non-Cooperative Games', *Annals of Mathematics*, **54**, pp. 286–95.

von Neumann, John and Morgenstern, Oskar (1944), *Theory of Games and Economic Behaviour*, Princeton, N.J.: Princeton University Press.

PART I

Preference Formation and Aggregation

2. Preferences, constraints and games: analysing Polish debt negotiations with international banks

Vinod K. Aggarwal and Pierre Allan*

In April 1991 the Paris Club of lender nations ratified a multilateral agreement to reduce Poland's official hard currency debt by 50 per cent. This momentous event prompted Janusz Sawicki, deputy minister of finance and Poland's chief debt negotiator to state:

> We hope that this agreement will be once and for all, since the essence of this agreement is the fact that we will now be able to service the contractual debt.[1]

For many banks, this agreement was a disaster. It meant that they would now be under severe pressure from Poland and creditor governments to offer Poland similar concessions on the debt owed them. But the banks have demonstrated that they are not willing to let Poland off so easily. According to one German banker:

> The Poles have yet to grasp how Western banking works or what write-offs mean for their creditworthiness. Simply put, if we loan you money and you don't pay it back, you never see another damn cent from our bank. That's just how it is.[2]

How will Poland fare in its current negotiations with private banks when it has been fairly unsuccessful during the rescheduling agreements of the 1980s? Based on a model that we discuss in this chapter, we examine rescheduling efforts during the 1980s and then suggest that Poland will probably prevail in the near future in its negotiations with private banks.

To examine Polish debt rescheduling, this chapter presents a framework to analyse international negotiations. Specifically, we use a situational theory of constraints to analyse constrained preferences of actors in different bargaining

* We are thankful for research funding from the Swiss National Science Foundation under grant no. 11.25552.88, and from the University of California Center for German and European Studies. For research assistance, we are indebted to Catherine L. Kuchta, Stephanie McLeod, and Brook Boyer and for comments to Cédric Dupont, Christian Schmidt and Marya Mogk.

contexts.[3] We identify the variables to specify these situational constraints by drawing from three levels of analysis: the international systemic level; the domestic politics level; and the decision-makers' level. By combining three variables, one from each of these levels, we are able to construct games of strategic interaction and predict the likely outcome of actors' bargaining efforts based on the well-known Nash solution concept.

Our theoretical approach is presented in section 1. Although we argue that Polish rescheduling with the banks can best be analysed in terms of five negotiating periods, we only provide a very detailed analysis of the first and last because of space constraints. Section 2 examines the period during 1981 in which the banks and Poland negotiated their first debt settlement. In section 3 we review rescheduling events in the interlude from the banks' first agreement until the late 1980s. We then provide our second in-depth analysis and study the current negotiations from 1990 onwards in section 4; there, we also analyse various scenarios to examine the future of Polish debt rescheduling with private banks. In conclusion, section 5 considers the utility of the approach we pursue here.

1 A THEORY OF CONSTRAINED PREFERENCE ORDERS AND GAMES[4]

For many years game theory has yielded insight into the problem of strategic interaction among actors. This approach is ideally suited to the study of international bargaining questions because it assumes that actors will make choices contingent upon their opponents' likely response. Unfortunately, most game theoretic studies neglect a more fundamental issue, namely: where do payoffs come from? That is, how should we determine the valuations that actors assign to different potential outcomes? Simply using the language of game theory to describe actors' interaction does not suffice. Our approach aims to work towards filling this gap by more systematically *deducing* payoffs based on the various situations in which actors might find themselves.

Another weakness in most game-theoretical analyses is that they simply posit a specific game to examine a given historical situation. Thus, studies often end up simply restating the empirical evidence using more complex terms – without having brought greater insight into the bargaining process. We show how our approach, while formal and deductive, avoids some of these problems and helps to illuminate the actual bargaining process.

In analysing strategic interaction, we must first specify actors' basic goals, the strategies available to them, and the potential outcomes resulting from their interaction. Later, we will see how three key dimensions of power resources are likely to constrain their *basic* preferences or objectives. We use the term constrained preference orders (CPOs) to refer to actors' valuations of different bargaining outcomes in the light of the constraints that they face.[5]

Although as a basic goal all actors would ideally prefer to win in any issue-specific bargaining game in which they are involved, they will be concerned about the implications of their actions on: (a) their domestic position; and (b) the larger overall strategic and economic concerns in which the issue-area is nested.[6] Thus, as we shall see, we shall need to include variables to tap these factors in our analysis.

To describe strategies and outcomes, we begin by specifying the options available to actors. Our primary focus is on predicting issue-specific outcomes; hence, we define actors' main strategic choices as 'accept' or 'demand' to represent their strategies in the issue-area. Subsequently, either actor may choose to retaliate against the other within the context of their larger relationship (abbreviated by three dots after the major choices). Turning to issue-specific outcomes, we label the combination of possible strategy choices in the four cells as 'mutual consensus', 'row prevails', 'column prevails' and 'no consensus'.

Figure 2.1 A general approach to define a strategic interaction situation

		Actor column	
		Accept and some retaliatory sequence (A . . .)	Demand and some retaliatory sequence (D . . .)
Actor row	Accept and some retaliatory sequence (A . . .)	Mutual consensus (MC) in issue area (cell 1)	Column prevails (CP) in issue area (cell 2)
	Demand and some retaliatory sequence (D . . .)	Row prevails (RP) in issue area (cell 3)	No consensus (NC) in issue area (cell 4)

1.1 The Three Situational Variables

We assume that three key situational variables related to different aspects of power will constrain actors' choices.[7] These are: an actor's overall power position; its issue-specific power; and the strength of its internal or domestic coalition. We see the issue-area as the key analytical entry-point for our analysis, because the outcomes we are centrally interested in predicting (for example, in Figure 2.1) are issue-specific outcomes of prevailing or consensus or the lack thereof. In any negotiation process, actors will manoeuvre to set the agenda for bargaining, which will partially be a subjective process.[8] The two other variables we use – overall power and the 'strength of coalitions' – refer to overall systemic concerns and domestic politics, respectively. Given this perspective, our variables can be seen as drawing from the three levels of analysis commonly used in international relations.[9]

In most international relations theory, the question of overall power tends to dominate analyses. This variable is central to 'Realist' analysis in world politics, which focuses primarily on the military power of nation-states. In our analysis, overall power is considered more broadly, encompassing economic and other variables; moreover, it need not apply only to nation-states but also to sub-national actors.

The issue-power variable is fundamental in recent analyses in international political economy. Scholars have focused on this variable, arguing that power may not always be fungible across issue-areas. Consequently, actors' resources in the individual issue-area (as initially defined by the bargaining parties) proves crucial for examining bargaining strength and outcomes.

Lastly, the focus on coalitional strength is an important element in the analysis of domestic politics and policy-making in general. As with the other variables, we do not restrict our analysis to a focus on state actors. Instead, we consider the coalitional stability of sub-national actors as well (for example, the unity of a coalition of banks in debt-rescheduling negotiations). In each case, we focus on constraints faced by the coalition's leader.[10]

If we dichotomize these three variables, we have the eight possible individual situations shown in Figure 2.2, in which each actor might find itself.

Figure 2.2: Actors' individual situations (IS)

	Overall resources	
Issue resources & stability	*Strong*	*Weak*
Issue strong & stable	IS 1	IS 5
Issue strong & unstable	IS 2	IS 6
Issue weak & stable	IS 3	IS 7
Issue weak & unstable	IS 4	IS 8

Note: 'IS *n*' refers to different individual situations reflecting different values of the three variables.

1.2 Constrained Preference Orderings and Games

We next turn to a specification of actors' ordinal preferences over the feasible outcomes in the light of the constraints they face.[11] To construct CPOs, we begin by postulating the likely effect of different values of the three variables on actors' objectives.[12] We specify four postulates stemming from the three variables to deduce a rank-order of actors' constrained preferences over outcomes.[13] Note that our postulates refer to preference orderings and not to strategic choices. The latter stem from the specific games in which actors find themselves.

We start with the following postulate concerning actors' orderings when one is issue strong or issue weak. For the sake of exposition, we take the point of view of actor 'Row' in the discussion that follows:

Postulate 1.1: RP is best when an actor is issue-strong.
Postulate 1.2: RP > NC or CP when an actor is issue-weak.

Issue-strong actors will be more willing to risk conflict than issue-weak actors because they think they can prevail and thus will always rank RP (row prevails) first; that is, RP is best for them. Issue-weak actors, by contrast, will be less aggressive and will not necessarily expect that prevailing (RP) is optimal for them.

We next specify the constraints and prospects arising from overall power resource considerations:

Postulate 2.1: NC > CP when overall strong.
Postulate 2.2: CP > NC when overall weak.

When Row is overall strong, its constrained preference is for a no-consensus outcome (NC) over having Column prevail (CP) The reasoning behind this postulate is as follows: actors who have overall strength are ready for a possible conflict escalation resulting from retaliatory moves rather than accept the other actor's issue-specific demands. Conversely, actors who are overall weak will generally be conciliatory in order to avoid confrontation and retaliatory measures. Thus, for them, Column prevailing (CP) in the issue-specific negotiations is likely to be their *constrained* preference over a no-consensus (NC) outcome.

The third postulate states that actors with a stable coalition will have a CPO for mutual consensus (MC) over no consensus (NC). By contrast, leaders of unstable coalitions will have constrained preferences for no consensus (NC) over mutual consensus (MC).

Postulate 3.1: MC > NC when coalitionally stable.
Postulate 3.2: NC > MC when coalitionally unstable.

Why might this be the case? First, a stable coalition provides its leader with room to discover whether both actors can be better off through integrative bargaining. His coalitional stability allows for the exploration of possible bargaining space.[14] Secondly, because the leader and his coalition have incumbency expectations, he is likely to have a longer time horizon, increasing the probability of the development of integrative bargaining solutions. Thirdly, leaders of stable coalitions will be better able to prevent free-riders who would benefit from the mutual consensus without joining the coalition. The converse of these arguments apply to leaders of unstable coalitions, who will be more willing to risk a no-consensus conflictive outcome (NC) over a mutual-consensus one (MC). In addition, leaders of unstable coalitions may be tempted to escalate conflicts through retaliatory measures to shore up support against real or imaginary external foes. This may allow them to bolster their coalition – albeit with some risk – and thereby increase their power.

The fourth postulate now examines the interactive effect of issue resources and domestic coalitions more directly:

> *Postulate 4.1*: MC > RP when an actor is issue-weak and stable.
> *Postulate 4.2*: RP > MC when an actor is issue-weak and unstable.

Stable issue-weak actors will have a constrained preference for a mutual consensus versus prevailing because they can envisage a consensus solution. For them, a mutual consensus outcome (MC) appears more realistic in the light of the constraints they face than attempting to prevail (RP).[15] An *unstable* issue-weak actor, by contrast, will have constrained preferences for prevailing (RP) over a mutual consensus (MC) outcome. Such actors will be less willing to seek a mutual consensus because their instability does not provide them with the necessary flexibility to find a consensual solution.

1.3 Deducing Full Constrained Preference Orders

We next combine these pair-wise preference comparisons to construct full constrained preference orderings for different individual situations. Figure 2.3 presents the CPOs from actor Row's perspective (for Column's view, simply replace RP by CP, and conversely everywhere).

Figure 2.3 shows that the four postulates generate unique preference orderings in five individual situations. In the three other cases, our deductive model predicts two different but relatively similar preference orderings. It is important to keep in mind that we have developed full constrained preference orderings by using our four simple postulates – not by specifying them directly from each individual situation.[16] In other words, our approach goes from simple pair-wise comparisons to more complex full preference orderings.[17]

Figure 2.3 The deduced constrained preference orderings from Row's perspective (for column, replace RP by CP and conversely)

Issue resources and stability	**Overall Resources** Strong *Pos. 2.1*: NC > CP	 Weak *Pos. 2.2*: CP > NC
Issue-strong *Pos. 1.1*: RP is best Stable *Pos. 3.1*: MC > NC	RP > MC > NC > CP Prisoner's Dilemma (IS 1)	MC RP > or > NC CP Chicken or Leader (IS 5)
Issue-strong *Pos. 1.1*: RP is best Unstable *Pos. 3.2*: NC > MC	MC RP > NC > or CP Deadlock or Deadlock analogue (IS 2)	RP > CP > NC > MC Hero (IS 6)
Issue-weak & stable *Pos. 4.1*: MC > RP *Pos. 1.2*: RP > NC or CP *Pos. 3.1*: MC > NC	MC > RP > NC > CP Stag Hunt (IS 3)	MC > RP > CP > NC Harmony (IS 7)
Issue-weak & unstable *Pos. 4.2*: RP > MC *Pos. 1.2*: RP > NC or CP *Pos. 3.2*: NC > MC	MC RP > NC > or CP Deadlock or Deadlock analogue (IS 4)	RP > CP > NC > MC Hero (IS 8)

Notes:
1. Game names refer only to symmetrical games formed by two similar preference orderings.
2. 'IS *n*' refers to different individual situations reflecting different values of the three variables.

1.4 Deriving Games from Full Ordinal Preferences

We now examine the specific games in normal form arising from various combinations of individual situations.[18] We assume complete information – in other words, that the actors involved in the bargaining are able to evaluate correctly their own and their opponent's preference orderings. The specific games are formed by combining the deduced single or double constrained preference ordering that results from an actor being in a certain individual situation. We examine the games that emerge from actors having perfectly symmetrical ordinal preference orderings (see Figure 2.4). Owing to space constraints, we only illustrate the possible deduced symmetric games since there are many fewer symmetric games than asymmetric ones. Outcomes are ordered from best (4) to worst (1). For all the individual situations, there are eight different symmetric games: Prisoner's Dilemma, Chicken, Leader, Deadlock, Deadlock analogue, Stag Hunt, Hero and Harmony.[19] In order to make predictions about the likely outcomes of a specific game, we use the standard Nash solution concept to find equilibria.

Most individual situation cells contain only one constrained preference ordering and therefore one game. In three cases we find two different orderings, giving rise to two symmetric games (IS 2, 4 and 5). However, these games are fairly similar. Deadlock analogue is like Deadlock: both actors' dominant strategy of demanding leads to an outcome of NC.[20] The game of Leader bears some similarities to Chicken. In such games, an asymmetric outcome is likely. Leader is often used to model situations of co-ordination, whereas Chicken is used to represent cases where one party imposes its will on the other through reckless behaviour. In the latter case, the asymmetry between parties is much greater, and the relative gain greater for the victorious party in Chicken than in Leader.

The variety presented here is only of the simplest kind – symmetrical games. In general, we would expect actors to be in different individual situations. Therefore our theory will often lead to asymmetrical games such as Called Bluff – resulting from a combination of one actor being in IS 1 (with Prisoner's Dilemma preferences) while the other is in IS 5 (with a preference ordering corresponding to Chicken).[21]

Examining all possible preference combinations produces 64 possible games.[22] We find that in each combination of individual situations, whether symmetrical or asymmetrical, the game or games resulting from that combination have the same strategy combinations in equilibria – that is, we predict the same *cell* outcomes. Put differently, even where we predict several possible games (because there are sometimes two possible CPOs for a given individual situation), in *all* cases we have the same cell solution(s). In six combinations of individual situations, there is no Nash equilibrium in pure strategies. Because our general approach does not allow for the greater precision of cardinal measurements of preference orderings, Nash solutions in terms of mixed strategies cannot be computed. Instead, we suggest that actors proceed with moves and countermoves, up to and until they end up in a Nash equilibrium where they will remain. They do so because only rarely do they act in their individual capacities – more often being leaders of coalitions and thus precluding a sophisticated strategy choice. Therefore, a series of moves and countermoves, or what we might term cycling, is our predicted outcome of all the games obtained in those cases.[23]

Figure 2.4 Deduced symmetric ordinal games

Issue resources and stability	Overall resources: Strong	Overall resources: Weak
Issue-strong & stable	Prisoner's Dilemma 3,3 / 1,4 4,1 / **2,2** (IS 1)	Chicken or Leader 3,3 / **2,4** 2,2 / **3,4** **4,2** / 1,1 **4,2** / 1,1 (IS 5)
Issue-strong & unstable	Deadlock or Deadlock analogue 2,2 / 1,4 1,1 / 2,4 4,1 / **3,3** 4,2 / **3,3** (IS 2)	Hero 1,1 / **3,4** **4,3** / 2,2 (IS 6)
Issue-weak & stable	Stag Hunt **4,4** / 1,3 3,1 / **2,2** (IS 3)	Harmony **4,4** / 2,3 3,2 / 1,1 (IS 7)
Issue-weak & unstable	Deadlock or Deadlock analogue 2,2 / 1,4 1,1 / 2,4 4,1 / **3,3** 4,2 / **3,3** (IS 4)	Hero 1,1 / **3,4** **4,3** / 2,2 (IS 8)

Note: Outcomes in bold are Nash equilibria.

When we combine individual situations, we never have more than four possible games. We find a unique game in 24 cases, two possible games in 30 cases, and four possible games in 10 cases. Considering the various outcomes, we find Row prevailing in 15 cases, 15 occurrences of Column prevailing, the no-consensus outcome in 15 instances, 15 cases of the twin outcomes of either Row *or* Column prevailing, six where we have cycling, three occurrences of the mutual-consensus outcome, and finally, one case of two Nash outcomes, either mutual or no consensus.[24]

This summary of our predicted games shows that our theory generates all possible cell equilibria. On the other hand, we obtain not one but two predicted outcomes in only 10 out of 64 cases.[25] Our theory is thus readily falsifiable in terms of the cell outcome predictions it makes.

In this paper, we use the model to analyse two of five periods of Polish debt negotiations from 1981 to the present. We also evaluate various scenarios and make predictions about the likely outcome of these negotiations in the years to come.

2 POLISH DEBT NEGOTIATIONS: 1981

Three principal actors have been playing a key role in Polish debt negotiations: Poland, Western creditor governments and commercial banks.[26] In many ways, the Polish negotiations with creditor governments are quite distinct from its bargaining with banks. In this chapter, we focus only on the game between the banks and Poland.

We define the issue-area as Polish debt-rescheduling negotiations beginning in March 1981, when Polish authorities announced their inability to continue debt servicing, to the present. In 1981 Poland informed Western banks of its inability to meet payments on loans coming due in the second quarter of that year.[27] Thus, our debt-rescheduling analysis focuses on events after Poland's March 1981 request for debt refinancing.

Following our game-strategic terminology, we delineate the different tactics pursued by both actors. Poland's option to 'accept' indicates its willingness to undertake adjustment policies, as well as to agree on higher interest rates and shorter maturity structures and grace periods. By contrast, a Polish 'demand' means that it insists on high concessions from the lenders. As for the lenders, they could agree to grant high concessions to the debtor (A), or simply insist on high adjustment (D).

By combining the actors' strategies, we obtain four possible outcomes (see Figure 2.5). If both actors basically accept the other's demand (MC), then the result would be a loan with the banks granting high concessions and Poland undertaking high adjustments. In the case of a no-consensus (NC) outcome, both actors insist that the other accept its demand and the result is the risk of default: the banks' insistence on high debtor adjustment would be confronted with tough resistance by Poland, who would accept only little or no adjustment. The Row prevails (RP) outcome means an asymmetrical situation favouring the banks, while the Column prevails (CP) outcome represents a 'win-like' outcome for Poland.

Figure 2.5 The Polish debt rescheduling game

		Poland	
		Accept . . .	Demand . . .
Banks	Accept . . .	High concessions High adjustments (MC)	High concessions Low/no adjustments Poland prevails (CP)
	Demand . . .	High adjustments Low/no concessions Banks prevail (RP)	Default possible or no agreement (NC)

The first period in this epoch of debt rescheduling between Poland and its Western creditor banks covers the year 1981. During this time one rescheduling agreement was concluded, after which Poland moved into a new individual situation and the second period began.

2.1 Applying the Theory to the Empirical Case and Deducing the Preference Orderings

In the first period of debt-rescheduling negotiations, we code Poland as coalitionally unstable, issue-weak and overall weak (IS 8). The banks were coalitionally unstable, issue-strong and overall weak (IS 6). We now discuss the rationale for these codings.

Poland

Overt challenges from Solidarity, divisions within the Communist Party, severe economic problems and unsettling Russian military activity in the region explain Poland's domestic instability during this period.

Beginning in late January 1981, Solidarity, the independent trade union movement, supported strike warnings with the potential to involve hundreds of thousands of workers.[28] In response, the Communist Party leader, Stanislaw Kania, accused Solidarity of promoting anarchy.[29] In early February 1981, in an effort to increase the military's influence in Poland, the Central Committee of the Polish Communist Party dismissed Prime Minister Jozef Pinkowski and nominated Defence Minister General Wojciech Jaruzelski to replace him.[30] A number of Western diplomats felt that the Polish situation had reached a dangerous stage and viewed the appointment of General Jaruzelski as the Warsaw government's last chance to restore order.[31]

Solidarity claimed that its members were being harassed, intimidated and beaten by security policemen.[32] Clear divisions existed within the Communist Party: the government newspaper *Zycie Warszawy* blamed the weakness and indecision among the Communist Party leaders for Solidarity's political and militant activities.[33] In addition to causing considerable domestic instability, this profound social strife directly affected the government's ability to implement effective economic reform.

Specifically, two principal barriers stymied Polish efforts to design an economic recovery programme: the uncertain debt repayment schedules, and the nation-wide political chaos.[34] Finally, on 13 December 1981, in an attempt to regain social and economic control, General Jaruzelski's military government invoked martial law. The General informed the West that this move was 'long overdue in light of the impending *coup d'état* by Solidarity'.[35]

Lastly, Soviet-led Warsaw Pact forces which began to manoeuvre in and around Poland in mid-March and continued to exert military and political pressure on Warsaw in early April, represented a further destabilizing force.[36] For these reasons we code Poland as unstable during 1981.

During this period Poland was weak in debt-related resources. Its international reserves of $0.5 billion in 1981[37] would not have been sufficient to service the outstanding private debt in convertible currency of $14.7 billion in 1981.[38] Moreover, Soviet exports to Poland had been falling in real terms more than Poland's exports to the USSR, symbolizing the Soviets' attempt to limit 'real' aid to Poland. Western nations severely curtailed import credits to Poland. Poland did receive, however, $1.4 billion for such purchases from agreements with Canada and France that had been secured before the imposition of martial law. Total new credits for 1981 were approximately $5 billion. Thus, Poland lacked the financial resources to service its foreign debt.

With respect to overall capabilities, Poland was weak during the first debt-rescheduling period. Western market access was crucial for Poland to obtain vital imports and to sell its exports for hard currency. Both gross imports from and exports to its four main creditors – Austria, France, West Germany, and the United States – decreased by 28 per cent between 1980 and 1981.[39] The lenders had virtually no investments or assets in Poland which the debtor could seize and use as leverage over its creditors.[40]

Banks

The lack of cohesiveness among the 19 members of the bank task-force, representing the 560 institutions involved, illustrates coalitional instability.[41] To discuss the banks' coalitional instability we examine both the organizational setting and group cohesiveness.[42] According to one source, the task-force acted as a channel of communication between its members and their national co-ordinating committees' countries, and each individual bank that was owed money by Poland:

> This cumbersome procedure is a far cry from the traditional rescheduling system where a steering committee of banks most heavily involved agrees terms with the borrower in closed negotiations.[43]

The American, Japanese and West European banks were divided on how to attack the issue. American banks, for example, preferred a 'wait and see' policy, while European banks (particularly those of West Germany) wanted a solution as soon as possible. In general, these attitudes reflected the varying exposures of the respective banks. West European banks' exposure totalled $15.8 billion in 1981, while US banks' lending amounted to only $1.2 billion in 1981.[44] In sum, although the banks were grouped behind a task-force, co-ordination of their various positions still proved difficult.

As for issue-specific resources, we code the banks as strong. Although overall approximately 50 per cent of Polish commercial loans were guaranteed by the creditor governments,[45] West German banks, the largest creditors, were not covered to the same extent as their American counterparts.[46] Nevertheless, as one source asserted, 'No German bank is in danger of collapsing because of the Polish loans . . . .'[47] This statement also held true for American banks. For example, according to Willard C. Butcher, then chairman of Chase Manhattan Bank, 'Our loans to Poland and less-developed countries are less than our risk was to well-known retailers several years ago.'[48]

Concerning overall-power capabilities, the banks were weak. Lending to East European countries involved somewhat more of a risk for ideological and political reasons. Banks, lacking sovereign power to enforce loan contracts, were hesitant to lend to Poland because of the nation-wide chaos and poor economic performance. Moreover, Poland was not a member of the IMF; thus banks could not require it to follow an IMF-prescribed stabilization program.

According to Figure 2.3 and taking Poland – issue weak, overall weak and unstable – as the 'Column actor', its constrained preference orderings are as follows (IS 8):

Poland's constrained preferences: CP > RP > NC > MC

Poland prefers a unilateral 'win' (CP) to the banks prevailing (RP), which is in turn preferred to default (NC). The least-preferred outcome for Poland is mutual consensus (MC).

The banks – issue strong, overall weak and unstable – were therefore in individual situation IS 6:

Banks' Constrained Preferences: RP > CP > NC > MC

They preferred an asymmetric 'win' (RP) to Poland prevailing (CP), which in turn was better than default (NC). The worst outcome for the banks was mutual consensus (MC).

2.2 The Deduced Game, Predicted Outcome and Empirical Testing

Using our game-strategic terminology, we delineate the different options available to both actors. Poland's option to 'accept' indicates its willingness to undertake adjustment policies as well as to agree on stiffer financing terms; on higher interest rates, for example. On the contrary, the Polish 'demand' means that it insists on high concessions from Western banks. As for the banks, they could agree to grant high concessions to the debtor (A), or simply insist on high adjustment (D).

Both Poland's and the banks' preference orderings correspond to a game of Hero. Figure 2.6 illustrates the resulting games from combining both actors' orderings.

Figure 2.6 Period 1 (1981) game

		Poland (Hero) Accept . . .	Poland (Hero) Demand. . .
Banks (Hero)	Accept . . .	High concessions High adjustments (MC) 1,1	High concessions Low/no adjustments Poland prevails (CP) **3,4**
	Demand . . .	High adjustments Low/no concessions Banks prevail (RP) **4,3**	Default possible or No agreement (NC) 2,2

Note: Bold figures indicate the Nash equilibria: the banks or Poland prevail.

If both actors accept the other's demand (MC), then the result would be an agreement, with the banks granting high concessions (such as new loans) and the debtor undertaking high adjustments. In the case of a no-consensus (NC) outcome, both actors insist that the other accept its demand and the result is a possible default. The banks, insisting on high debtor adjustment, would be confronted with tough resistance by Poland, whose strategy is to accept little or no adjustment. The Row prevails (RP) outcome would signify a 'win' for the banks, whereas the Column prevails outcome would represent a unilateral 'victory' for Poland (CP).

Thus, the game in the first period could either produce an asymmetric outcome favouring the banks (RP) or one favouring Poland (CP). Next we examine the bargaining to compare our predicted outcome of (RP) or (CP) with the actual results of the negotiations.

2.3 Negotiations and Actual Outcomes

Empirically, Poland and the banks did not agree on a debt settlement because of Poland's failure to come up with the necessary goodwill payments. The banks prevailed, however, and refused to reschedule without receiving the required funds (RP).

Negotiations

Following Poland's request to refinance nearly $3 billion of debt falling due in 1981, Western banks reluctantly began to negotiate with Poland. In late April, members of the task force met with Polish authorities to discuss arrangements for rescheduling. Although Poland originally asked for all 1981 commercial maturities, amounting to $3.1 billion, to be rescheduled, banks were only willing to discuss the remaining three-quarters due.[49]

A significant disagreement arose between the banks over whether or not Poland's first-quarter payment of $0.8 billion should be included in the agreement.[50] In addition, the banks' insistence that Poland 'produce much more detailed economic information as well as a workable economic stabilization program before [the banks would] agree to rescheduling Polish debt' further impeded negotiations.[51]

The coalitional differences within the 19-bank task-force caused negotiations to proceed slowly. Two American banks were at the centre of this problem. Chase Manhattan Bank argued that their two loans totalling $0.5 billion for Poland's copper industry should not be part of a rescheduling agreement.[52] As for Citibank, it received a relatively large payment on its principal after the effective moratorium began on 26 March. Citibank officials preferred that this undisclosed sum be deducted from the amount to be rescheduled.[53]

In late June, the banks reached an agreement on their negotiating stance. The task force recommended that all but 5 per cent of the $2.4 billion owed in 1981 be rescheduled for 7.5 years with an interest rate of 1.75 per cent over LIBOR and a 1 per cent penalty charge.[54] In addition to these provisions, the task force demanded that Warsaw report detailed economic information to its creditor banks, including the status of its debts owed to the Eastern Bloc.[55]

The task force met again on 22–23 July in Zurich to reconfirm its position *vis-à-vis* the debtor and to submit the proposals to Poland. In response, Polish government officials expressed their disappointment over the direction of negotiations and called the bankers' offer 'too tough'.[56] Besides seeking to reschedule 100 per cent of the principal, the Poles wanted to decrease the interest rate by as much as ¼ per cent. In fact, on 27 July, Zygmunt Krolak, a senior Polish government official, charged the American, French and West German banks with being 'unco-operative'.[57]

On 21 August, Polish officials agreed on the 22 July terms for the long-term debt falling due in 1981. Before the banks would sign the agreement, they insisted on receiving the remaining 1981 interest and the 5 per cent principal not covered in the agreement. Combined, these sums amounted to roughly $0.5 billion before the end of December 1981.

Although the prospects for signing the agreement before the 28 December deadline looked promising, Poland's economy continued to weaken. Three days after the outbreak of martial law, 13 December 1981, Polish authorities asked the Western banks for a short-term loan of $0.3 billion to help cover the interest due. The banks, however, rejected Poland's request and the agreement was postponed, as the Warsaw government refused to pay its upcoming debt service obligations.

Outcome

In sum, the outcome was one of low/no concessions by the banks, and high adjustments by Poland, that is, Row prevails (RP). The bankers refused to reschedule before Poland repaid 100 per cent of the interest and 5 per cent of the principal on its 1981 obligations, and they denied the debtor any bridge financing. Poland, on the other hand, repaid the 5 per cent principal (even though it wished to reschedule 100 per cent of the principal), and the 1981 interest before receiving any new funds or rescheduling the 1981 debt.

Furthermore, Poland introduced two reform initiatives during 1981. The first, which was designed by then Prime Minister Pinkowski, began on 1 January 1981. Its major objectives were to axe new industrial projects and redirect its supplies into housing development, rationalize the price system, limit imports (except food), decentralize economic management, and restructure wages to eliminate unjustifiably high incomes.[58] On 12 February newly appointed Prime Minister General Jaruzelski announced his own, inward-looking, economic reform programme.[59] The General accused Solidarity of impeding economic progress and said that the main thrust of the reform, a 110 per cent price increase, would be implemented during the last quarter of 1981.[60] Hence, the results are consistent with one of our predicted outcomes of the banks prevailing (RP).

3 POLISH DEBT NEGOTIATIONS: 1982–90

The first rescheduling period was followed by three others. Each time the constraints facing one of the actors changed, a new individual situation resulted, often leading into a new preference ordering over the outcomes. First, while the banks continued to face the same constraints, Poland, due to the imposition of martial law in December 1981, became coalitionally stable. This marked a second period which only lasted until late in 1982, when the banks were able to unify too, thus leading to a third period of negotiations. In 1985, however, Poland once again became unstable, a situation that lasted until 1990.

Given these momentous changes, one would expect new constrained preference orderings to arise and new games to form. In effect, this happened. However, despite these new individual situations and constraints, our predicted outcomes basically remain the same: in periods 2 and 3 we expect the banks to prevail; in period 4 we predict that either the banks or Poland should prevail. The empirical results confirm these theoretical predictions: throughout the period from 1982 until 1990, the banks always prevailed. Let us now turn to a discussion of these three periods.

3.1 Period 2: 1982

As mentioned above, the situational change marking the second period of debt negotiations was Poland's return to political stability. In December 1981 the declaration of martial law coupled with the imprisonment and/or detainment of approximately 4000 Solidarity supporters, including its leader Lech Walesa, successfully blunted the challenge from Solidarity and restored order in Poland.[61] Hopes for a legal revival of the symbolic Solidarity labour union movement were quashed with the government ban on all labour unions, announced on 8 October 1982.[62] By late 1982, a Solidarity-orchestrated strike had failed to materialize, the government had freed Lech Walesa and had patched up relations with the Roman Catholic Church. Thus, General Jaruzelski's stature both inside and outside the Communist Party was 'stronger than ever'.[63] In fact, Jaruzelski's regime then suspended martial law from 30 December 1982. The domestic climate in Poland had been 'normalized.'

With Poland's new-found stability, its constrained preference ordering came to resemble that of an actor in the game of 'Harmony,' whereas the banks remained in 'Hero', with the predicted outcome of a clear success for the banks (see Appendix 2.2).

Two agreements were signed in 1982. In the first round, both actors reached an accord in which the terms were highly favourable to the creditors. On 6 April 1982, representatives of the Warsaw government and the 19-bank task-force signed an agreement rescheduling 95 per cent of the 1981 principal and interest, amounting to $2.4 billion, over a period of 7.5 years with an interest rate of 1.75 per cent above LIBOR, and charged a 1 per cent handling fee.

Following this, the banks realized the pressing need to begin rescheduling the nearly $3.4 billion owed to them in 1982. Despite a worsening of the Polish economy, private creditors were reluctant to offer fresh financing without a guarantee from their respective governments. Poland requested to be released from paying the interest on the 1982 obligations. The banks flatly refused this proposal, but failed to reach a consensus on extending new credits.[64]

The negotiations were pursued during the summer, and finally, on 3 November 1982, a rescheduling agreement was signed representing another success for the private creditors. The banks refused to provide any new financing and to capitalise any of the interest, but they agreed to extend repayments of the $1.1 billion in interest until March 1983, and to recycle 50 per cent of this as trade credits. Poland, for its part, agreed to repay 5 per cent of the principal by November 1983. The remaining 95 per cent would be repaid over $7\frac{1}{2}$ years rather than over ten years as Poland had requested. Moreover, in July 1982 Poland began to open its economy to foreign investment in order to stimulate the private sector.[65]

In conclusion, the banks prevailed in both negotiating rounds, offering low/no concessions while Poland made high adjustments. Poland was never offered any fresh financing and rescheduled its obligations annually, whereas Latin American debtors, during the same time period, rescheduled their debts on better terms.[66] In sum, our prediction of the banks prevailing, an (RP) outcome, is consistent with the actual results in both negotiation rounds.

3.2 Period 3: Late-1982–5

The banks resolved their differences and became stable (moving from IS 6 to IS 5) during this period, which commenced in late 1982 and concluded in early 1985. There were no changes in Poland's individual situation; it remained coalitionally stable, issue-specific weak and overall weak (IS 7).

The banks, unstable in the first period, stabilized in the second period, as no major splits appeared between them. At the beginning of the rescheduling process in 1981 and 1982, the banks faced many obstacles to co-operation. By the end of 1982, however, the larger banks used various methods to ensure the co-operation of smaller banks. For example, when smaller banks threatened to break the ranks and call for a default, the larger banks either bought them out or told them to 'shut up'.[67]

As in 1982, negotiations in this period between the commercial banks and Poland took place in two rounds. In both rounds the banks prevailed (RP) as our theory again predicts – though there are two new games in this instance: the banks with a 'Leader-' or 'Chicken-like' preference ordering, and Poland in 'Harmony.'

The first round of negotiations began in mid-March 1983. The final agreement was signed on 3 November 1983 based on the following terms: $1.5 billion in principal would be rescheduled over ten years with a five years' grace period, repayment of the interest in three instalments by the end of 1983, with 65 per cent of that recycled in the form of short-term trade credits. This was an undeniable triumph for the banks because they simultaneously refused Poland's request for fresh funding and increased the interest rate by ¼ per cent higher than that of the two previous agreements. Poland's terms were much more onerous than those of Argentina, Brazil and Mexico, in which the private creditors granted $4.2 billion, $4.8 billion and 5.4 billion, respectively, in new funding during 1983. Moreover, Poland still settled its debts on an annual basis while Argentina and Mexico rescheduled their obligations on a multiyear basis during this time.[68]

The second agreement was signed on 13 July 1984. It rescheduled approximately $1.6 billion in loans due between 1984–7.[69] The sum was to be repaid over ten years with a five-year grace period with an interest rate of 1⅝ per cent over LIBOR. Poland received $0.6 billion in extra-trade credits, $0.3 billion of which was new trade credits towards the end of 1984, and $0.3 billion was an extension or a recycling of existing trade credits. Once again the banks triumphed by making low/no concessions in rescheduling Poland's debts, while Poland made high adjustments. The amount of fresh money involved paled in comparison to Poland's needs and to the sums that were offered to other Latin American debtors during this time. For example, in 1984 Argentina, Mexico and Brazil received $2.4 billion, $3.8 billion, and more than $6.5 billion, respectively, in new money.[70] Also, these countries rescheduled 100 per cent of the principal on their 1983 and 1984 obligations, whereas Poland resettled 95 per cent.

3.3 Period 4: 1985–90

During 1985 Poland became unstable again, moving from individual situation IS 7 to IS 8. This change produced a new period of negotiation during which two rescheduling agreements and one interim accord were signed. We now analyse briefly the cause of Poland's renewed domestic instability.

Following the murder of pro-Solidarity Father Popielski in October 1984, which incited national protests and international outrage, conflict mounted. The murder also spurred an 'about face' in the pro-government stance of Cardinal Glemp, the Catholic leader.[71] Despite its lack of legal status, the Solidarity movement continued to thrive through underground efforts. In late 1985 General Jaruzelski warned his comrades that they faced a long struggle against enemies inside the country.[72] But Solidarity had become a system of values and a way of life which could not be arrested.[73] The Poles had not been 'Czeched'.[74]

With Poland in individual situation IS 8 and the banks remaining in IS 5, the theory predicts two possible outcomes: either the banks or Poland will prevail. Empirically, the banks continued prevailing in the three rounds of this period.

In the first round in June 1986, the private creditors slightly decreased the interest rate on the rescheduling; they also reduced the amount of the down payment. Nevertheless, the banks only agreed to restructure these obligations a second time, the first having been in 1981–2, and refused to grant Poland the new financing that it desperately needed. Warsaw, for its part, dropped its request for new money. This was a major concession at the time not only because Latin American debtors had been receiving new funds throughout the early 1980s but also because Argentina was granted more than $0.9 billion in 1986 and Mexico was given close to $14 billion the same year.[75]

In the second round, negotiators finally reached an agreement on 20 July 1988. The terms were very similar to the ones of those of the year before. Although the banks agreed to lower the interest rate slightly and reschedule Poland's debt on a multiyear basis, they refused to offer Poland new financing before an IMF-endorsed programme was in place. Mexico, on the other hand, was awarded $3.5 billion in new money in 1988,[76] before it settled its arrears of more than $2 billion with the Western creditor governments.

By mid-June 1990, in the third round of negotiations, Poland agreed to pay the $24 million in capital two months earlier. Warsaw expected the banks to heed the creditor governments' request for debt forgiveness. The banks, however, balked at this suggestion and acceded only to an extension of the repayment period. The banks had won again.

Following the first debt rescheduling agreement between the Polish government and the Western private creditors in 1981, negotiations thus continued throughout the decade while the state of Poland's economy worsened. Throughout all four debt negotiating periods, the coalition of banks extracted high concessions from Poland despite changes in the actors' constraints and individual situations. What rescheduling results will the 1990s bring? Poland has gained strength in both issue and overall power resources, while the banks' coalition has become unstable again. Will the new constraints facing both actors lead to different results? Let us now turn to events in the 1990s and examine the situations.

4 POLISH DEBT NEGOTIATIONS: 1990–PRESENT

4.1 Applying the Theory to the Empirical Case and Deducing the Preference Orderings

In this period, Poland remained unstable but became both issue and overall strong, which puts it into IS 2. The banks were still issue strong and overall weak but became unstable, as in period 1 (IS 6).

Poland

While the June 1989 parliamentary elections signalled the collapse of the Communist regime with the subsequent election of Tadeusz Mazowiecki,[77] the first non-Communist prime minister, a formidable political challenge overshadowed the positive aspects of Poland's transition to democracy: the reconciliation of the various internal factions of Solidarity and the need to secure popular support for Minister of Finance Leszek Balcerowicz's ambitious programme for economic reform.

Despite Mazowiecki's achievements in 'laying the foundations for a democratic political system and a functioning market economy',[78] the thirteen months of recession Poland weathered under his government eroded his broad-based popular support. In addition,

> for Solidarity, the unexpected transition from opposition to government has revealed political strains within Solidarity and the economic sacrifices demanded by the Balcerowicz Plan underlined the need for a strong, popularly elected government.[79]

In the end, Mazowiecki's lack of flexibility, combined with an inability to share power, accept criticism, or even sell his own governments' success to the people, determined his poor showing in the November 1990 elections.[80] Lech Walesa also recognized these divisions within the Solidarity movement. His personal political ambitions and the desire to rebuild Solidarity's political force inspired his campaign for the presidential elections. Despite such efforts, Poland remained 'a country of many contradictions and divisions',[81] and the first completely free elections since the Second World War reflected these internal struggles.

In the first round of the presidential elections,[82] three candidates dominated: Prime Minister Mazowiecki; Lech Walesa, the leader of Solidarity; and Stanislaw Tyminski, a Polish-Canadian businessman running on a populist platform.

Tyminski was completely unknown in Poland a month prior to the elections, when he collected the 100,000 signatures required to participate.[83] The constituency's positive reaction to the challenge of Tyminski revealed their clear dissatisfaction with the other two mainstream candidates. One quotation aptly summarizes the general attitude toward the upcoming elections:

> The real winner will be democracy, yet from every corner come apprehensive voices; the election threatened stability, and Mr Walesa, as president will threaten democracy.[84]

Lech Walesa, the front-runner of the first round of Poland's presidential elections with 40 per cent of the votes, tried to unite Solidarity against the increasing threat from Tyminski. Despite Walesa's 75 per cent victory in December, the first nationwide free elections provoked serious political divisions: the Solidarity alliance was severed and a multiparty structure arose. Under Walesa a remarkable shift in power from the government and Parliament to the President occurred, creating additional political instability.

The continued transition and political unrest only aggravated Poland's economic problems. As labour unrest mounted in the first months of 1991, Walesa threatened to use 'all the means at his disposal',[85] including force or rule by decree, to keep potential anarchy under control. In addition, two economic factors contributed directly to Poland's domestic instability: its enormous outstanding debt and the obvious implications of rigorous economic reform on wages. As noted by one journalist,

> keeping real wages low during a period of capital accumulation and new company formation is sure to erode political consensus behind reform as income differentials and gaps between social classes widen.[86]

In sum, Poland's attempt at large-scale political transformation during a period of considerable economic duress produced significant domestic instability.

However, Poland did move into a position of issue and overall strength. Two principal reasons justify the coding of issue strength: an increase in hard currency reserves and improved access to alternative sources of funding from institutions such as the IMF and the World Bank.

Poland's hard-currency reserves grew by $1.7 billion in the first quarter of 1990.[87] This leap contributed to Poland's increase in convertible currencies from a small surplus of $100 million in 1989 to $2.2 billion in 1990.[88] In addition, the World Bank offered to create an international fund to promote industrial restructuring in Poland by speeding up the inflow of private-sector foreign investment.[89] This idea was in response to the low level of foreign investment generated by Poland's radical economic reform in 1990. In late 1990 the IMF backed a $1 billion fund to stabilize the zloty. The OECD developed this programme which was ultimately co-ordinated by the European Economic Commission. In addition, the IMF organized a financial package including SDRs worth about $2.5 billion. These trends toward increases in Poland's reserves and increased potential to obtain fresh financing explain its transition to a position of issue strength.

Concerning overall capabilities, Poland became overall strong. Its transition to democracy and to a free market economy mark this period. This fact, combined with a relatively low dependence on foreign trade and negotiations to improve access to other markets, explain Poland's move to a position of overall strength. Poland's level of self-sufficiency also increased its overall power capabilities. For example, with exports comprising only 15 per cent of total GNP in 1990, Poland's dependence on foreign trade was much lower than that of countries such as France (22 per cent), West Germany (33 per cent), and The Netherlands (55 per cent).[90] In the first quarter of 1991 overall export performance was strong; between January 1990 and January 1991 Poland's merchandise exports in dollars increased by 59 per cent.[91] In March 1991, the EC accepted the 'principle of asymmetry': it would open its markets to imports from Poland more than Poland would to exports from the EC. Lastly, Poland's transition to democracy increased its overall strength *vis-à-vis* Western creditor governments because Poland would now be able to link debt relief to its political transformation.

Banks

Following almost eight years of success in maintaining internal cohesion, the banks split once again in 1990 over the delicate issue of debt reduction. However, Poland's June 1990 proposal to both banks and Western creditor governments to reduce its outstanding debt by 80 per cent did not come as a surprise. Nevertheless, internal co-ordination over how to respond to this proposal proved extremely difficult for the banks.

Distinct differences developed between the ten main creditor banks led by Barclays of the United Kingdom. Some banks insisted, and continue to insist, upon payment of part of Poland's outstanding interest arrears before beginning comprehensive debt-restructuring talks. Other banks, from countries including Germany and Japan,[92] adamantly opposed any form of debt reduction, preferring temporary reduction of interest payments.

Another division between the banks emerged over the general approach to continued debt rescheduling. One group, including the Bank of America and many others, simply wanted to 'get rid of the problem'; a second group felt that they could 'sit it out' and were thus willing to be more conciliatory.[93] This latter group included the French banks and Dresdner Bank (the most exposed German bank).

Increased loan-loss reserves by banks throughout the 1980s had a dual effect: this provisioning increased the banks' issue strength; but diverging bank interests produced further internal divisions. For example, varying country tax codes and different investment maturity structures in Poland (German and French banks had long-term investment in Poland compared to Switzerland with only short-term interests) caused banks to divide over how to proceed with Polish rescheduling. Thus the bankers' coalition became unstable.

The banks remained issue-strong. Three main factors contributed to the banks' issue-strength: sufficient levels of loan-loss reserves, Poland's renewed IMF membership, and the April 1991 agreement with the Paris Club. The banks' continuous efforts to augment loan-loss reserves throughout the 1980s explains their ever-increasing issue-strength during the same period. By successfully accumulating sufficient loan-loss reserves, the banks are presently in a position to sit tight and wait; in the words of one banker, 'we have time, that's an asset'.[94]

Poland's readmission to the IMF in 1986 has also increased the banks' issue strength during the last two periods. Poland is now obliged to disclose vital economic information and allow IMF teams to evaluate its economy to propose programmes for reform. Thus, the likelihood that the banks will be repaid improves, if indirectly, with the implementation of an IMF 'conditionality' programme and the new money that accompanies such agreements.

It is possible that the Paris Club agreement which reduced Polish debt by 50 per cent could increase the banks' strength. Banks could argue that, with a 50 per cent debt reduction, Poland now has more funds available to repay them. Earlier, Polish analysts worried that concluding an agreement with the Paris Club prior to any talks with the banks would not only cause Polish debt prices to surge on the secondary market but also prompt banks to insist that Poland 'accept a high benchmark price to determine the discount used under the Brady Plan'.[95] Following this logic, banks would also be less likely to agree to any debt reduction schemes. In the opinion of advisers to the Polish Ministry of Finance, for debt reduction to be successful, creditor governments must 'press banks to accept a big enough reduction'.[96]

Finally, banks remain overall weak in this period as they have been throughout the entire epoch. In addition, the Paris Club agreement of April 1991 further limits the banks' overall strength. The fact that the creditor governments have already agreed to debt reduction limits the banks' ability to appeal to their respective governments as potential allies. By offering their support for debt reduction, the governments have already allied with Poland. The banks, in general, have come under considerable pressure from creditor governments not only to proceed with debt restructuring but also to agree to terms at least as favourable as those of the April Paris Club agreement.

4.2 The Deduced Game, Predicted Outcome, and Empirical Testing

In the negotiations between Poland and the banks in the current period no agreement has been reached. In this case our deductive model allows us to predict that if Poland is unstable, issue strong and overall strong (IS 2) and the banks are unstable, issue strong and overall weak (IS 6), we would then witness an outcome in which Poland prevails (see Figure 2.7) – whatever the specific game they are in.

Figure 2.7 Period 5 (1990–present) games

		Poland (Deadlock) Accept . . .	Demand . . .
Banks (Hero)	Accept . . .	High concessions High adjustments (MC) 1,2	High concessions Low/no adjustments Poland prevails (CP) **3,4**
	Demand . . .	High adjustments Low/no concessions Banks prevail (RP) 4,1	Default possible or No agreement (NC) 2,3

		Poland (Deadlock Analogue) Accept . . .	Demand . . .
Banks (Hero)	Accept . . .	High concessions High adjustments (MC) 1,1	High concessions Low/no adjustments Poland prevails (CP) **3,4**
	Demand . . .	High adjustments Low/no concessions Banks prevail (RP) 4,2	Default possible or No agreement (NC) 2,3

Note: Bold figures indicate the Nash equilibria: the banks prevail.

The simple fact that massive debt reduction is on the agenda for negotiations in this period indicates that so far Poland has been prevailing. At this point we feel that this situation represents the most plausible scenario. The asymmetrical game resulting from the combination of Deadlock-type preferences for Poland and those of Hero for the banks produces the Nash equilibrium of Column prevailing (CP) – the asymmetrical outcome favouring Poland. We now turn to a discussion of the negotiations.

4.3 Negotiations and Actual Outcomes

Negotiations between Poland and the commercial banks have taken a back seat in favour of the negotiations and ultimate Paris Club agreement between Poland and the creditor governments in April 1991. Because the governments were Poland's largest creditors, the banks initially approved of a Paris Club rescheduling before their own negotiations with Poland. However, the final Paris Club decision on interest capitalization poses problems for the banks who were less willing to negotiate similar terms.[97]

The Paris Club and the London Club have met only twice. The whole group of bankers met with Paris Club people in January 1990 and for a second time in early 1991. At this time banks made it clear that they would seek a different agreement. In the agreement that Poland negotiated with the Paris Club, banks must give Poland 'comparable terms'.[98] According to one Swiss banker, the banks' response to Polish demands for an agreement with terms similar to those of the Paris Club agreement was simply: 'go renegotiate your agreement with the Paris Club'.[99] In other words, given their distinct interests, the banks are not willing to adopt a negotiating stance *vis-à-vis* Poland similar to that pursued by the creditor governments.

However, bank creditors who originally asked Poland to pay 15 per cent of the interest due on its outstanding debt, fell in line with the creditor governments and postponed interest payments until the end of March 1991. In the short run this move took pressure off Poland, but at the cost of raising total debt.[100] This incident, plus what appears to be the banks' tacit acceptance of the principle of debt reduction, supports our model's prediction that Poland will prevail.

In March 1991 Poland's foreign bank creditors, led by Barclays Bank, met in Paris. Talks on restructuring Poland's more than $10 billion in bank debt are progressing slowly, partly because the two sides have not been able to agree on the treatment of unpaid interest arrears. For their part, the banks still want Poland to pay 30 per cent of interest arrears before discussing a general proposal on the rest of the debt.[101]

Janusz Sawicki, Deputy Finance Minister and Poland's chief debt negotiator, commented that he expected the banks

> to be reasonable. We have never had a reasonable word from the banks. We have not had an opportunity to negotiate with the banks. They say that they first want money up front to clear up past due interest of about $1 billion.[102]

Sawicki insisted that the banks must accept a deal at least as generous as that negotiated with the Paris Club, simply because Poland has consistently treated the banks better than the creditor governments. If banks attempt to profit from the Paris Club agreement to try to force further payment, Sawicki's views are clear: 'They can get as tough as they want, but in that case they will get nothing.'[103]

In May 1991, however, Poland shifted its negotiating position. Poland was now willing to negotiate a settlement of its roughly $1 billion in interest arrears owed to commercial banks, provided that the banks agreed to follow this minor concession with talks on a comprehensive debt agreement – a relatively major concession. To make its offer concrete, at the beginning of May 1991 Poland extended a $100 million interest payment and a deposit of 20 per cent of interest to go forward into a special account at BIS.[104] According to bankers, this offer opened up debt negotiations to include discussions for a comprehensive debt agreement, 'including debt-cut options where loans will be exchanged for concessional bonds'.[105]

Despite progress made between the two parties, in June 1991 Poland announced its intentions to 'delay scheduled repayments and prolong negotiations with commercial creditors'.[106] Later in June two days of negotiations over a potential debt-reduction package for Poland failed to produce any concrete results. The main hurdle remains: the banks want an agreement on outstanding interest arrears prior to commencing discussions over reduction in principal.

Since then talks have not been able to get off the ground for numerous reasons. A late-summer debt scandal involving Polish debt officials wreaked havoc in Warsaw government circles. On 23rd August 1991 Grzegorz Zemek, Head of the FOZZ (set up in 1989 to administer Poland's debt), was arrested for mismanagement of government funds. According to reports, Polish officials accused Zemek of using the funds of private Polish companies to buy back Polish debt on the secondary market. Not only have such activities led to 'enormous losses by the [Polish] treasury', but a 1988 rescheduling agreement with the Western creditor banks forbids buy-back operations of this type.[107] The scandal reached large proportions when Manufacturers Hanover Trust and Dillon Reed were cited for channelling the funds.[108] Moreover, Polish Prime Minister Bielecki dismissed Janusz Sawicki, Poland's chief debt negotiator and FOZZ chairman, the day after Zemek's arrest.

Hopeful signs emerged at the beginning of September 1991 when Polish officials announced the resumption of debt talks between Prime Minister Bielecki and Jeffry Stockley of the London Club of Western banks. But a breakthrough continued to be hampered as Poland insisted that the banks match write-off terms granted by the Paris Club.

By May 1992 Stockley reported that negotiations would not resume until Poland struck a deal with the IMF. The Fund stated that an agreement was likely by the end of July, but thus far both sides have yet to come to terms.[109]

4.4 Scenario Analyses from 1993 Onward

Unless there is a situational change, we predict that Poland will prevail. Given the fact that an agreement has yet to be reached between the banks (in IS 6) and Poland (in IS 2), our model can also be used to predict what outcomes would result from possible plausible deviations from our base-line coding of the actors in joint situation 62 (IS 6 and IS 2).

It appears that the banks will remain issue-strong and overall weak, and we envisage that Poland, with its transition to democracy, will remain overall strong. The possible changes are, then, the following: either the banks or Poland could regain their internal stability or Poland could return to a position of issue-weakness if economic reforms fail to produce adequate foreign currency reserves or if new sources of credit dry up.

The combination of these three possible changes (banks become stable, Poland becomes stable, Poland becomes issue weak) produces eight possible outcomes (see Figure 2.8).

In six cases out of the eight, we predict that Poland will prevail. Only in the two situations in which Poland is coalitionally stable, issue-weak and overall strong do we predict an outcome of cycling around the game matrix – since no Nash equilibrium in pure strategies exists.

The results are interesting in that if Poland becomes stable as well as issue weak, then its chances of securing a favourable agreement are reduced since this produces an outcome other than a Polish victory – cycling. Otherwise there is no change in the outcome whether the banks are unstable (as we coded them up to now) or stable.

In conclusion, given the respective individual situations of Poland and the banks, our model predicts the present negotiations over private-debt rescheduling or reduction to produce an outcome in which Poland prevails. We also examine seven additional scenarios based upon plausible changes in the individual situation of either actor. According to the model, in six out of the eight possible cases, Poland should prevail. As an agreement has yet to be negotiated between Poland and Western banks, we can only offer our prediction and expose our model to this future test.

Figure 2.8 Scenario analysis of Polish private debt negotiations, from 1993 onwards

Banks	Poland	Joint situation	Predicted outcome
Issue strong Overall weak Unstable	Issue-strong Overall strong Unstable	62	Poland prevails
Issue strong Overall weak Unstable	Issue strong Overall strong Stable	61	Poland prevails
Issue strong Overall weak Unstable	Issue weak Overall strong, Unstable	64	Poland prevails
Issue strong Overall weak Unstable	Issue weak Overall strong Stable	63	Cycling
Issue strong Overall weak Stable	Issue strong Overall strong Unstable	52	Poland prevails
Issue strong Overall weak Stable	Issue strong Overall strong Stable	51	Poland prevails
Issue strong Overall weak Stable	Issue weak Overall strong Unstable	54	Poland prevails
Issue strong Overall weak Stable	Issue weak Overall strong Stable	53	Cycling

5 CONCLUSION

The future of Polish debt rescheduling remains in doubt. However, based on a model which combines domestic political and economic variables, this chapter has argued that we can gain insight into the international bargaining process and make contingent predictions of debt-rescheduling outcomes.

We have sought to go beyond the standard use of game theory to analyse international strategic interaction. We have shown that payoffs need not simply be assumed but rather that some theory of the origin of payoffs can be a first step in any analysis. The predictions we make of games are based on a deductive model. We do *not* look at the actual outcomes of negotiations to construct payoffs, a weakness in most other approaches using game theory.

We began by presenting our theoretical approach to deduce constrained preferences from postulates about actors' goals and three key 'situational variables' that are likely to constrain and influence these goals. These are issue-specific resources, overall resources, and domestic or internal coalitional stability. Games of strategic interaction result from a combination of these deduced game preferences. Our theory can be validated empirically, as we demonstrated by examining all the cases of Polish debt rescheduling during the 1980s, starting from the first in 1981. In all instances, the predictions made by the model were borne out by the historical evidence.

Such an approach circumvents some of the problems facing analysts of contemporary history. The data needs are easier to satisfy with our approach because we do not require data about decision-makers' preferences. The preferences are obtained directly from an observation of the situation in which actors find themselves. Our model has minimal data requirements for two reasons: we only need very rough (that is, dichotomized) measures of three types of power capabilities and we need not make assumptions about cardinality of preferences.

Finally, our deductive theory allows us to make conditional predictions. For the current period, we argued that the changes in the banks' stability and in Poland's issue- and overall strength will allow Poland eventually to prevail. Currently (early 1993), this appears to be the direction that events are taking. We have suggested some alternative scenarios of what might happen if sharp changes take place in either Poland's or the banks' individual situations. In almost all the cases which appear to be reasonable extrapolations into the future, we expect Poland to prevail.*

* *Postscript.* On 11 March 1994, as we predicted, Poland secured an exceptionally generous commercial bank debt agreement calling for more than a 40% reduction in principal and an unprecedented reduction of 20% in interest arrears. Under the Brady Plan, from 1989 to the present, large Latin American debtors have received an average 35% reduction in principal and no reduction in interest arrears. For details see *New York Times*, 12 March 1994.

APPENDIX 2.1 POLAND'S EXTERNAL DEBT, 1970–90 (BILLION US DOLLARS)

	Gross external (1)	CMEA convertible (2)	Total Non-CMEA convertible (3)	Paris Club (4)	London Club: private banks (5)
1970	1.2				
1971	1.1				
1972	1.2				
1973	2.6				
1974	5.2				
1975	8.4				
1976	12.1				
1977	14.9				
1978	18.6				
1979	23.7				15.0
1980	24.1				15.1
1981	25.9	2.0	23.9	9.2	14.7
1982	26.3	1.9	24.4	11.0	13.4
1983	26.4	1.8	24.6	13.3	11.3
1984	26.9	2.2	24.7	15.8	8.9
1985	29.3	2.3	27.0	16.8	10.2
1986	33.5	2.5	31.0	20.0	11.0
1987	39.2	2.5	36.7	24.4	12.3
1988	39.2	2.4	36.8	26.2	10.6
1989	41.4	2.3	39.1	28.8	10.3
1990	42.5	2.2	40.3	29.3	11.0

Notes

(1) The 1970–88 figures are from Economic Commission for Europe (1990, Appendix Table C11). The 1989–90 figures are from Antowska-Bartosiewicz and Malecki (1991, Table 1). All 1990 figures in the table are for the first half of 1990, except those of the private banks, which represent the amount as of September 1990.

(2) Figures from 1981–7 are from Table E6, staff compilation from data supplied by the Polish authorities. The 1990 figure is from Antowska and Malecki (1991, Table 5). The 1988 and 1989 figures are interpolated.

(3) The 1981–7 figures are column 1 minus column 2. The 1990 figure is from Antowska-Bartosiewicz and Malecki (1991, Table 5).

(4) The Paris Club debt is the difference between the non-CMEA convertible debt minus the private debt.

(5) All commercial debt figures are from BIS *Quarterly Reports*, 1979–91.

APPENDIX 2.2: POLISH DEBT NEGOTIATIONS WITH WESTERN BANKS: 1981–PRESENT

Period 1: 1981

Banks: issue strong, overall weak, coalitionally unstable (IS.6)
Banks' preferences: RP > CP > NC > MC.

Poland: issue weak, overall weak, coalitionally unstable (IS 8)
Poland preferences: CP > RP > NC > MC.

Game:

		Poland: Hero	
		Accept	Demand
Banks: Hero	Accept	1,1	**3,4**
	Demand	**4,3**	2,2

Game outcome: Banks or Poland prevail
Empirical outcome:
Round 1 (March 1981–December 1981): Banks prevail

Period 2: 1982

Change: Poland becomes stable

Banks: same as in period 1

Poland: issue weak, overall weak, coalitionally stable (IS 7)
Poland preferences: MC > CP > RP > NC.

Game:

		Poland: Harmony	
		Accept	Demand
Banks: Hero	Accept	1,4	3,3
	Demand	**4,2**	2,1

Game outcome: Banks prevail

Empirical outcomes:
Round 1 (January 1982–April 1982): Banks prevail
Round 2 (May 1982–November 1982): Banks prevail

Period 3: late 1982–5

Change: Banks Become stable

Banks: issue strong, overall weak, coalitionally stable (IS 5)
Banks' preferences: RP > MC or CP > NC.

Poland: same as in period 2

Game:

		Poland: Harmony				
		Accept	Demand		Accept	Demand
Banks: Leader	Accept	2,4	3,3	Chicken	3,4	2,3
	Demand	**4,2**	1,1		**4,2**	1,1

Game outcome: Banks prevail

Empirical outcomes:
Round 1 (March 1983–November 1983): Banks prevail
Round 2 (March 1984–July 1984): Banks prevail

Period 4: 1985–90

Change: Poland becomes unstable

Banks: same as in period 3

Poland: issue weak, overall weak, coalitionally unstable (IS 8)
Poland preferences: CP > RP > NC > MC.

Game:

		Poland: Hero				
		Accept	Demand		Accept	Demand
Banks: Leader	Accept	2,1	**3,4**	Chicken	3,1	**2,4**
	Demand	**4,3**	1,2		**4,3**	1,2

Game outcome: Banks or Poland prevail

Empirical outcome:
Round 1 (November 1985–September 1986): Banks prevail
Round 2 (April 1987–August 1988): Banks prevail
Round 3 (June 1989–January 1990): Banks prevail

Period 5: 1990–present

Change: Banks become unstable and Poland becomes issue and overall strong

Banks: issue strong, overall weak, coalitionally unstable (IS 6)

Banks' preferences: RP > CP > NC > MC.

Poland: issue strong, overall strong, coalitionally unstable (IS 2)
Poland preferences: CP > NC > MC or RP.

Game:

		Poland: Deadlock		Poland: Deadlock Analogue	
		Accept	Demand	Accept	Demand
Banks: Hero	Accept	1,2	**3,4**	1,1	**3,4**
	Demand	4,1	2,3	4,2	2,3

Game outcome: Poland prevails

Empirical outcome: pending

Summary: Polish Private Debt Negotiations

Time period	Individual situation (IS) Banks	Individual situation (IS) Poland	Theoretical predictions	Empirical outcome
1981	IS 6 (Hero)	IS 8 (Hero)	Banks or Poland prevail	Banks prevail
1982	IS 6 (Hero)	IS 7 (Harmony)	Banks prevail	Banks prevail
1982–85	IS 5 (Chicken or Leader)	IS 7 (Harmony)	Banks prevail	Banks prevail
1985–90	IS 5 (Chicken or Leader)	IS 8 (Hero)	Banks or Poland prevail	Banks prevail
1990–present	IS 6 (Hero)	IS 2 (Deadlock or D'lock analogue)	Poland prevails	Pending

NOTES

1. *Financial Times*, 3 May 1991.
2. *The New York Times*, 20 February 1992.
3. This 'situational' approach was first developed by Aggarwal (1989 and forthcoming) for the analysis of international debt rescheduling. A generalization of this model and ordinal formulation of payoffs can be found in Aggarwal and Allan (1991) and Aggarwal and Allan (1994).
4. This section is adapted from Aggarwal and Allan (1992 and 1993).
5. See Aggarwal and Allan (1991 and 1994) for a detailed discussion of terminology. The term 'constrained preference order' was suggested by Robert Powell.
6. For a discussion of nesting, see Aggarwal (1985), pp. 27–8.
7. For a more detailed discussion of these variables see Aggarwal (1989 and forthcoming).
8. See Keohane and Nye (1977, pp. 64–5) for a discussion of the subjective nature of issue-areas.
9. Although, at some level, all three variables can be seen to be 'cognitively' defined, once the issue-area has been agreed upon for purposes of negotiations, we can examine each actor's power capabilities along the lines of an issue-structural analysis. Naturally, an important part of the long-run bargaining process is for weaker parties to redefine the negotiation arena to their advantage, either through knowledge-based appeals or through simple power-based linkage strategies. But in the short run, we can examine each actor's 'objective' issue-area capabilities relevant to the specific bargaining problem at hand. We should note that all issue-area analyses face this problem of subjective and objective considerations.
10. All three variables are conceptualized as measuring power in absolute and not relative terms. In other words, both actors might be powerful on all dimensions, and both might be simultaneously lacking in power resources.
11. We could also specify *cardinal* preference orderings. One approach to this would be through the construction of utility equations focusing on actors' goals and their weighing of these goals. Unfortunately, cardinality requires greater precision in empirical coding and additional strong assumptions about actors' behaviour. On the other hand, specification of cardinal preferences may sometimes allow us to distinguish between two apparently plausible equilibria and provide us with greater empirical insight. See Aggarwal (1989 and forthcoming) for examples of this.
12. For an earlier approach to deduce ordinal preference orderings and games based on goals and changing power configurations, see Allan (1983).
13. The postulates that follow are almost identical to but somewhat stricter than the ones presented in Aggarwal and Allan (1991).
14. This does not imply, of course, that actors with a stable coalition are more willing to acknowledge other actors' demands than to make their own demands. It only indicates that stable coalitions permit rational and serious discussion of the other actor's demands, thereby increasing the perceived chances of arriving at a mutually advantageous solution.
15. Only for such actors do we find the prevailing outcome (RP) to be the second-best solution, not the best.
16. We also postulate that preferences are transitive, that is, consistent. The transitivity rule is usually used for defining rationality. Note also that we do not consider ties in constructing the preference orderings for expository purposes.
17. The advantage of this deductive method is three-fold. First, it is much easier to justify pairwise comparisons of outcomes than actors' constrained preferences across the whole set of outcomes. Secondly, the logic of building up full preferences from simpler postulates is transparent, allowing for critical analysis of our assumptions. Thirdly, this approach allows us to ensure consistency for preference orders across different individual situations since the logic used in constructing full preferences is identical.
18. We wish to develop a theory which is quite general. Therefore, the normal-form game representation is more appropriate than alternative game-theoretical formalizations such as sequential games with imperfect information, for two reasons. First, the specific sequence

of retaliatory moves and countermoves will vary from case to case, and thus is difficult to model in general. Moreover, as noted, our modelling objective is to predict issue-specific outcomes, and not the specific pattern of retaliation that might follow. Secondly, empirically, introducing incomplete information may neither always be realistic nor very useful because of the severe coding problems that this introduces. For a discussion of the advantages and disadvantages of modelling choices with respect to information conditions and game forms, see Aggarwal and Dupont (1992).

19. For a discussion of Prisoner's Dilemma, Chicken, Leader, Hero and Deadlock, see Snyder and Diesing (1977). Jervis (1978) discusses Stag Hunt, and Harmony is discussed at length by Keohane (1984).
20. Downs *et al* (1985, p. 121), use the term 'type' to refer to games that are almost identical in structure and play. A better term appears to be an 'analogue' game.
21. See Snyder and Diesing (1977) for a discussion of Called Bluff.
22. 8 × 8 CPOs, since we distinguish between actors Row and Column (RP different from CP).
23. Joint situations JS 3&5, 5&3, 3&6, 6&3, 3&8 and 8&3. See Aggarwal and Allan (1993) for empirical examples of such a case.
24. This occurs in Stag Hunt (both actors in IS 3). If we want to define a unique equilibrium solution, we can use the solution concept 'in the strict sense' since one of those Nash solutions (MC = 4,4) is Pareto-superior to the other one (NC = 2,2); see Luce and Raiffa (1957, p. 107). For an elaborate treatment of the criteria for choosing among Nash equilibria, see Harsanyi and Selten (1988).
25. In other cases we have looked at, we have found unique equilibria and such findings are supported empirically. See Aggarwal and Allan (1992 and 1994).
26. The Council for Mutual Economic Assistance (CMEA), with the Soviet Union holding about 85 per cent of the loans (see Appendix 2.1) represents another Polish creditor. This study does not analyze this relationship for two principal reasons. First, debts among COMECON nations have been recorded in convertible roubles. Secondly, trade credits and methods of barter play large roles in restructuring or settling these debts. Both these facts make gathering and interpreting actual data difficult.
27. *The Times,* 14 April 1981.
28. The *New York Times,* 23 January 1981.
29. Ibid., 4 February 1981.
30. Ibid., 10 February 1981.
31. Ibid., 11 February 1981. In the months directly following the imposition of martial law, most Western countries debated 'whether martial law had been justified by the prevailing circumstances', and the conditions for its repeal *(Keesing's Contemporary Archives,* 30 April 1982, p. 31453).
32. *New York Times,* 8 March 1981.
33. Ibid., 15 March 1981.
34. *International Herald Tribune,* 29 April 1981.
35. *The Financial Times,* 14 December 1981.
36. *New York Times,* 9 April 1981. Already in 1980 the West had focused their attention on the concentration of Warsaw Pact troops in the region of Poland *(Keesing's Contemporary Archives,* 20 February 1981, p. 30721).
37. World Bank (1990, p. 314).
38. Data supplied from BIS, *International Banking and Financial Market Developments, Quarterly Reports,* 1979–91. Although the annual net sums of Poland's private debt declined slightly from 1981 to 1982, this was because of the two reschedulings (April 1982 and November 1982) and the fluctuations in hard-currency exchange rates during this period.
39. Gross imports from these four major lenders declined from \$3.4 billion in 1980 to \$2.5 billion in 1981. Similarly, gross exports to these four nations dropped from \$2.7 billion in 1980 to \$1.9 billion in 1981 *(The Direction of Trade Statistics Yearbook,* 1990, pp. 328–9).
40. In fact, joint ventures were not legal in Poland until 1986.
41. It is quite typical for creditors to be coalitionally unstable during the onset of rescheduling agreements. One source puts it clearly: 'First the rescheduling process itself is a mess. At the start of the negotiations almost everything is up for grabs – not only the debts that are to

be included in the rescheduling, but also the terms and the conditions of the final agreement with the borrower. It is ironic that the banks which co-operated so easily in making the original loans through the smooth functioning of the syndicated-loan market should find it so hard to co-operate in picking up the pieces when those loans turn sour' (*The Banker*, March 1982).

42. On this point, see Aggarwal (forthcoming).
43. *The Financial Times*, 22 June 1981.
44. See McCarthy (1988, pp. 230–31) BIS–Federal Reserve Bank and Country Exposure Lending Survey, US Federal Reserve Bank. By the end of August 1981, the European banks' exposure was broken down as follows: West German ($4.05b.), French ($2.58b.), Austrian ($1.82b.), British ($1.75b.) and Italian ($1.09b.) (*The Economist*, 13 February 1982).
45. *The Times*, 14 April 1981.
46. At the close of 1981, West German banks were owed about $4.5 billion; $2.7 billion, representing 60% of the total, was not guaranteed by the government. *The Times*, 9 December 1981.
47. *Wall Street Journal*, 24 December 1981.
48. Ibid., 21 December 1981.
49. *International Herald Tribune*, 29 April 1981.
50. Ibid.
51. *The Financial Times*, 7 June 1981.
52. *International Herald Tribune*, 22 May 1981.
53. Ibid.
54. Ibid., 26 June 1981.
55. Ibid.
56. Ibid.
57. *Quarterly Economic Review of Poland*, 3 (1981), p. 15.
58. *Keesing's Contemporary Archives*, 20 February 1981, p. 30720.
59. Ibid., 3 April 1981, pp. 30797–8.
60. Ibid., 2 October 1981, p. 31112.
61. Reactions against the Communist regime manifested themselves in various forms, including spontaneous demonstrations in Warsaw, Gdansk and Poznan, the sudden appearance of hundreds of clandestine publications and Solidarity radio broadcasts, a boycott of the junta by the elite of the nation, the public denunciation of 'collaborators', and the creation of an illegal 'provisional commission' to co-ordinate independent union movements (*International Herald Tribune*, 12 May 1982).
62. Party leader Jaruzelski implied that the recent domestic social upheavals could set back the lifting of martial law, which had been promised in his July speech (*New York Times*, 10 October 1982).
63. *The Financial Times*, 3 December 1982.
64. Banks strongly opposed rescheduling interest and principal for fear of setting an unwanted precedent (*New York Times*, 6 May 1982). Rescheduling of principal is normally done on the condition that interest payments are met.
65. *Keesing's Contemporary Archives*, 8 October 1982, p. 31736.
66. Mexico settled its obligations on a multiyear basis (from 1982–4) over a twelve-year time period and with the benefit of $2.5 billion in fresh funding; Argentina was offered $1.8 billion in 1982 at a lower rate of interest (1⅝ per cent over LIBOR), and Brazil was granted more than $3 billion in new money the same year.
67. Comment made by J. Paul Horne, first Vice-President of Smith Barney Harris Upham, Inc., in Paris (quoted from *International Money Management*, 23 August 1982).
68. Aggarwal (1987, Appendices A-1, A-2 and A-3).
69. The four-year debt deal did not signify that commercial banks had confidence in the future of Poland's economy; rather, it indicated a recognition by these banks that Warsaw's debt was not worth annual haggling. After the 1984 settlement, the bankers informed the Poles that they would be receiving new short-term credit in 1984 (*The Financial Times*, 30 April 1984).
70. See Aggarwal (1987, Appendices A-l, A-2, and A-3).
71. Glemp's conciliatory attitude towards the government angered many Church followers. They pointed to the glaring contrast between his platform and that of the former, much-loved

and more militant, Cardinal Wyszynski (*The Economist,* 12 January 1985). On 18 June 1985 Jaruzelski met Cardinal Glemp for the first time since January 1984; however, this is believed to have been more of a pre-election manoeuvre than a bona fide effort to establish normal, friendly relations with the Roman Catholic Church. In fact, the Polish government refused to define the legal status of the Church.

72. The *New York Times,* 29 November 1985.
73. *The Economist,* 7 June 1986.
74. Ibid., 3 August 1985.
75. See Aggarwal (1987, Appendices A-l and A-3).
76. Aggarwal (forthcoming).
77. To many Poles the 4 June 1989 elections for the Sejm and the Senate represented the 'end of communism in Poland' (*East Europe Reporter,* 4 (1), (Winter 1989)). These elections, however, were not entirely free, as the Communist Party was guaranteed 65 per cent of the seats in the lower house of Parliament and that the post of president would be given to General Wojciech Jaruzelski (*The Financial Times,* 22 November 1990).
78. *Report on Eastern Europe,* 2 (1), (4 January 1991).
79. *The Financial Times,* 20 November 1990.
80. Ibid.
81. Ibid., 27 November 1990.
82. The Polish presidential elections of November–December 1990 proceeded in two rounds because no candidate achieved a majority in the first round. The split between the top three of the six candidates was as follows: Lech Walesa 40%, Stanislaw Tyminski 23%, Tadeusz Mazowiecki 18% (*The Financial Times,* 27 November 1990). Mazowiecki withdrew from the race and resigned from his position as Prime Minister prior to the second round of the elections scheduled for 9 December 1990. In this run-off between Walesa and Tyminski, 50% of the population voted, and Walesa emerged victorious with a 75% majority.
83. *The Financial Times,* 23 November 1990.
84. *New York Times,* 8 December 1990.
85. *The Financial Times,* 14 June 1991.
86. Ibid., 20 November 1990.
87. Ibid., 1 June 1990.
88. Ibid., 3 May 1991.
89. Ibid., 25 October 1990.
90. Ibid., 3 May 1991.
91. Ibid.
92. In some countries the governments exert substantial influence over the banks to alter their negotiating position. For example, the Japanese bankers take the following approach: 'Once the government tells me what to do, I do it' (*Interview with Swiss Bankers,* 17 June 1991). On the other hand, in Switzerland there is practically no government interference.
93. Ibid.
94. Ibid.
95. Antowska-Bartosiewicz and Malecki (1991, p. 58).
96. Ibid.
97. *Interview with Swiss bankers,* 17 June 1991.
98. Ibid.
99. Ibid.
100. *The Financial Times,* 20 November 1990.
101. Ibid., 17 March 1991.
102. Ibid., 3 May 1991.
103. Ibid.
104. Ibid.
105. Ibid., 13 May 1991.
106. *The Economist,* 15 June 1991.
107. *The Financial Times,* 23 August 1991.
108. Ibid. and 11 October 1991.
109. Ibid., 13 May 1992.

REFERENCES

Aggarwal, Vinod K. (1985), *Liberal Protectionism: The International Politics of Organized Textile Trade,* Berkeley, Cal.: University of California Press.

—— (1987), *International Debt Threat: Bargaining Among Creditors and Debtors in the 1980s,* Policy Papers in International Affairs, no. 29, Berkeley, Cal.: Institute of International Studies.

—— (1989), 'Interpreting the History of Mexico's External Debt Crises', in Barry Eichengreen and Peter Lindert (eds), *A Long Run Perspective on the Debt Crisis,* Cambridge, Mass.: MIT Press.

—— (forthcoming), *Debt Games: Strategic Interaction in International Debt Rescheduling,* New York: Cambridge University Press.

Aggarwal, Vinod K. and Allan, Pierre (1991), 'Obiettivi, Preferenzie, e Giochi: Verso una Teoria della Contrattazione Internazionale' [Goals, Preferences and Games: Towards a Theory of International Bargaining], in Paolo Guerrieri and Pier Carlo Padoan (eds) *Politiche Economiche Nazionale e Regimi Internazionali,* Milan: Franco Angeli.

—— (1992), 'Cold War Endgames', in Pierre Allan and Kjell Goldmann (eds), *The End of the Cold War: Evaluating Theories of International Relations,* Dordrecht: Nijhoff.

—— (1993), 'Cycling over Berlin: the Deadline and Wall Crises', in Dan Caldwell and Timothy J. McKeown (eds), *Diplomacy, Force, and Leadership: Essays in Honor of Alexander L. George,* Boulder, Col.: Westview.

—— (1994), 'The Origin of Games: a Theory of the Formation of Ordinal Preferences and Games', in Michael Intriligator and Urs Luterbacher (eds), *Cooperative Models in International Relations Research,* Norwell, Mass.: Kluwer Academic Publishers, in press.

Aggarwal, Vinod K. and Dupont, Cédric (1992), 'Modeling International Debt Rescheduling: Choosing Game Theoretic Representations and Deriving Payoffs', paper presented at the American Political Science Association meetings, Chicago, September.

Allan, Pierre (1983), *Crisis Bargaining and the Arms Race: A Theoretical Model.* Cambridge, Mass.: Ballinger.

Antowska-Bartosiewicz, Iwona and Malecki, Witold (1991), *Poland's External Debt Problem,* Warsaw: Friedrich-Ebert Foundation Poland, Economic and Social Policy Paper 1.

Baldwin, David (1985), *Economic Statecraft,* Princeton, N.J.: Princeton University Press.

Bank of International Settlements (BIS), International Banking and Financial Market Developments (1991), *Bank of International Settlements Quarterly Reports.*

The Banker, various issues.

Brams, Steven J. (1985), *Superpower Games: Applying Game Theory to Superpower Conflict,* New Haven, Conn.: Yale University Press.

Cohen, Benjamin (1986), *In Whose Interest?: International Banking and American Foreign Policy,* New Haven, Conn.: Yale University Press.

Downs, George, Rocke, David and Siverson, Randolph (1985), 'Arms Races and Cooperation', *World Politics,* **38**.

East Europe Reporter, various issues.

Economic Commission for Europe (1990), *Economic Survey in Europe in 1989–1990; Eastern Europe and the Soviet Union: Balance of Payments in Convertible Currency,* prepared by the Secretariat of the Economic Commission for Europe, Geneva and New York: United Nations .

The Economist, various issues.

Euromoney, various issues.

The Financial Times, various issues.

Harsanyi, John and Selten, Reinhard (1988), *A General Theory of Equilibrium Selection in Games,* Cambridge, Mass.: MIT Press.

International Herald Tribune, various issues.

International Monetary Fund (IMF) (1990), *The Direction of Trade Statistics Yearbook, 1990,* Washington, D.C.: IMF.

—— (1991), *IMF Survey,* Washington, D.C.: IMF.

International Money Management (1982).

Interview with Professor Witold Malecki, 19 April 1991, in Warsaw.

Interview with Professor Stanislaw Raczkowski, 18 April 1991, in Warsaw.

Interview with Swiss Bankers, 17 June 1991, in Switzerland.

Keesing's Contemporary Archives, various issues.

Keesing's Record of World Events, various issues.

Keohane, Robert (1984), *After Hegemony: Cooperation and Discord in the World Political Economy,* Princeton, N.J.: Princeton University Press.

—— and Nye, Joseph (1977), *Power and Interdependence,* Boston Mass.: Little, Brown.

Krasner, Stephen D. (ed.) (1977), *International Regimes,* Ithaca, N.Y.: Cornell University Press.

Luce, Duncan and Raiffa, Howard (1957), *Games and Decisions,* New York: John Wiley.

March, James (1966), 'The Power of Power', in David Easton (ed.), *Varieties of Political Analysis,* Englewood Cliffs, N.J.: Prentice-Hall, pp. 39–70.

—— and Simon, Herbert (1948), *Organizations,* New York: John Wiley.

McCarthy, Paul I. (1988), 'Polands's Long Road Back to Creditworthiness', in Paul Marer and Siwinski Wlodzimierz (eds), *Creditworthiness and Reform in Poland,* Bloomington, Ind.: Indiana University Press.

New York Times, various issues.

Quarterly Economic Review of Poland, various issues.

Rapoport, Anatol and Guyer, Melvin (1966), 'A Taxonomy of 2 × 2 Games', *General Systems,* **11.**

Report of Economists on meeting with Polish officials in Warsaw (1983).

Simon, Herbert (1945), *Administrative Behavior: A Study of Decision-making Processes in Administrative Organizations,* New York: Free Press.

Snyder, Glenn H. and Diesing, Paul (1977), *Conflict among Nations: Bargaining, Decision-making, and System Structure in International Crises,* Princeton, N.J.: Princeton University Press.

Taylor, Michael (1976), *Anarchy and Cooperation,* New York: John Wiley; Rev. edn (1987), *The Possibility of Cooperation,* Cambridge: Cambridge University Press.

The Times, various issues.

United States Senate (1983), *The Polish Debt Crisis*: Hearings before a Subcommittee of the Committee on Appropriations, Washington, D.C.: United States Government Printing Office.

Wagner, Harrison (1983), 'The Theory of Games and the Problem of International Cooperation', *American Political Science Review,* **70.**

Wall Street Journal, various issues.

World Bank (1990), *World Debt Table, 1989–1990: External Debt of Developing Countries,* Washington, D.C.: World Bank.

World Bank *et al.* (1988), *External Debt: Definition, Statistical Coverage and Methodology*, a Report by an International Working Group on External Debt Statistics of the World Bank, International Monetary Fund, Bank of International Settlements, and Organization for Economic Co-operation and Development, Paris: World Bank.

3. Unpacking the national interest: an analysis of preference aggregation in ordinal games

Lars-Erik Cederman*

INTRODUCTION

In order to avoid Arrow's paradox, rational choice modellers often resort to the unitary-actor assumption. However, this assumption ignores important aspects of foreign policy-making in pluralist societies. This chapter presents a simple model that suggests how to reconcile majority voting with an analysis of the national interest, here conceptualized as a transitive social preference order. It is shown that imposing certain 'substantively rational' restrictions on individual-level preferences automatically leads to a single-peaked preference profile. A simple, normal-form game illustrates how the aggregated preference orders can be used in a strategic setting.

The term the 'national interest' is among the most abused in politics. Politicians do not hesitate to justify their acts by referring to this elusive concept. Furthermore, great confusion reigns about the scientific value of the term (see George and Keohane, 1980). But what is the national interest? In today's complex pluralist societies, it is not even obvious that it exists. Things were less complicated for the Machiavellian prince. Until the French Revolution, it was generally assumed that the absolute ruler incarnated the national interest. Today, as a wave of democratization sweeps through the world, the 'privileged' group of dictators is rapidly shrinking. Alexander George and Robert Keohane (1980, p. 219) put it aptly:

> With the 'democratization' of nationalism . . . the relative simplicity of the concept 'raison d'état' was eroded, and the state itself came to be seen as composed of

* An earlier version of this article was presented at the Inaugural Pan-European Conference of the ECPR Standing Group on International Relations, Heidelberg, Germany, 16–20 September 1992. I should like to thank the participants of this panel as well as Chris Achen, Douglas Dion and Simon Hug for their extremely helpful comments. All errors and misinterpretations are, of course, my own. The financial support from the American–Scandinavian Foundation and the Program for Peace and Security at the Social Science Research Council/MacArthur Foundation is gratefully acknowledged.

different interests. In the era of liberal democracy, 'l'état, c'est moi' was no longer an acceptable answer to the question of sovereign legitimacy. The national interest came to reflect a weighting of various diverse interests . . . as different groups within the policy competed to claim it as a legitimizing symbol for their interests and aspirations, which might by no means be shared by many of their compatriots.

These complications have also strongly influenced the scientific debate. The central issue is whether the state can be treated as a 'unitary rational actor'. Following Graham Allison's (1971) classic study of the Cuban Missile Crisis, many scholars, especially the critics of rationalist paradigm, have been inclined to equate rational choice models and the unitary-actor assumption. The present chapter takes issue with this view. In contrast, it argues that the link between the rational-choice paradigm and the unitary-actor perspective can and should be broken. I present a model of foreign policy-making that makes it possible to define the national interest precisely as a consistent social preference order. Moreover, it shows under which conditions the unitary-actor assumption is justified. The result suggests that Arrow's paradox may be less crippling than it was thought to be.

By restricting the attention to a class of simple but substantively important situations, I show that the national interest can be said to exist. Instead of imposing technical assumptions about single-peakedness of the preferences, I show that empirically relevant restrictions pertaining to 'substantive rationality' in conflictual situations automatically satisfy single-peakedness. Consequently, this finding removes one of the theoretical excuses for not opening up the 'black box' of domestic politics.

Nevertheless, it is very important to realize that this conclusion is subject to important qualifications. Single-peakedness does not follow if the number of policy choices increases. In such cases, it is necessary to impose even stronger constraints on the preference orders. However, by encouraging a formal derivation of the conditions for the existence of a collective utility function, this article suggests that similar investigations may be fruitful.

The chapter is organized as follows: section 1 reviews the relevant formal literature; section 2 introduces the individual level assumptions of the model; section 3 introduces the aggregation process from which the major theoretical result is derived; in section 4, the framework is applied to two simple strategic games; finally, in the conclusion, the findings are summarized and some caveats and possible extensions are discussed.

1 THE CURSES OF AND THE CURES FOR ARROW'S THEOREM

When Allison (1971) wrote his book, the unitary actor assumption and the rational choice literature were thought to be inseparable. Deterred by the impressive findings of Arrow (1963), decision theorists generally refrained from disaggregating the decision-making process. Treating the nation-state as a 'billiard ball', these authors often justified the unitary actor assumption by arguing that (a) a single leader dominates the foreign policy-making of states, or (b) that there is such a strong consensus on important foreign policy issues that a well-defined 'national interest' exists (cf. Bueno de Mesquita, 1981; Krasner, 1978).[1]

It does not take much imagination to find important counterarguments to these two rationales. First, although only one individual has the formal responsibility for foreign policy decisions, this person is generally not insensitive to domestic pressures. For instance, Achen (1988, p. 6) warns that 'if presidents are simply influenced by a variety of other actors, their decisions need not be identical with those of any one actor's personal preferences. Hence, they need not be representable.' Secondly, despite allegations of the opposite (Bueno de Mesquita, 1981), the domestic debate on foreign affairs is not necessarily characterized by unanimous consent.[2] Even in countries with a traditionally strong executive, there is a strong trend toward the fragmentation of foreign policy-making: 'The bipartisan consensus that allowed presidents to control American foreign policy from Pearl Harbor to the early stages of the Vietnam War gave way to ideological polarization' (Mann, 1990, p. 2).[3]

Therefore, it seems that critics of rational choice are right in urging the rationalists to pay more attention to 'the wisdom of old adages like the camel's being a horse designed by a committee' (Ferguson and Mansbach, 1991, p. 373). Despite the eagerness of the rationalists to promote the gospel of methodological individualism, their applications in international relations almost without exception treat the nation-state as a unitary actor, thus smuggling in a holistic perspective through the back door. Strangely enough, the quest for methodological individualism never reaches below the state level. Reification of the state limits the individualistic perspective to states whereas the true individuals remain black-boxed.

Fortunately, Allison's crude 'Model I' no longer reflects the state of the art of rational-choice theory (Bendor and Hammond, 1992). Recent advances in formal theory have enabled analysts to include the disaggregation of political units as an integrated part of the research agenda. Drawing on information economics, one type of model represents the nation-state as a principal-agent pair (for example, Bueno de Mesquita and Lalman, 1990, 1992; Richards *et al.*, 1992). While highly useful in the study of the strategic relations between the government and its constituents, these models involve such small numbers of actors that they fall short of capturing the interaction among the multitude of actors present in pluralistic political systems. Christopher Achen (1988) presents a different approach which does not suffer from this limitation. Based on techniques from duopoly theory, his framework features a strategic situation in which the behaviour of several domestic influences and a central decision-maker is representable by a continuous utility function.

Most of these models rely on cardinal representation of the actors' preferences. Under that assumption, it becomes possible to evade the spell of Arrow's impossibility theorem (Sen, 1982, ch. 11; see also Achen, 1988). However, many situations in foreign policy-making lend themselves more naturally to an ordinal utility representation. From a practical perspective, the often scanty data makes the creation of cardinal indicators very hard. In other words, it would be desirable to study the aggregation process in such a framework, despite the difficulties imposed by Arrow's theorem.

To see why this logical obstacle is worth taking seriously, we briefly outline the conditions for the theorem's applicability as well as some of the remedies that have been suggested. The impossibility theorem enumerates a number of desirable properties of a social choice rule (or a social welfare function in Arrow's terminology) and then concludes that there is no such rule that satisfies these conditions. The conditions are:

U the universal domain constraint
P the strong Pareto condition
I Independence of irrelevant alternatives
R Transitivity
D absence of a Dictator.

The U condition implies that the social-choice rule produces a ranking for all choices and all sets of individuals with arbitrary preferences. In particular, with fewer than three actors or fewer than three alternatives, it is easy to find a rule that satisfies the conditions. The P condition means that if everybody prefers an alternative, this alternative is collectively preferred by the society. The I condition is perhaps somewhat less obvious, but nevertheless compelling: adding an alternative to a set of alternatives should not affect the social ranking of the previous options. The R condition captures a narrow, but important notion of rationality; it requires that there be a consistent social ranking of alternatives. In this chapter, I shall identify the national interest with the existence of a transitive preference order.[4] Finally, the D condition excludes the possibility of a dictator. Of course, the presence of an actor with complete personal control over the national interest automatically violates the majority rule.

Several 'cures' have been suggested to escape from the 'curse' of the impossibility theorem. Arrow himself offered a number of ideas on this theme (see also Luce and Raiffa, 1957, pp. 340–57). Provided that the basic formulation of the problem remains the same (such as ordinal utility and so on), the analyst typically relaxes one or several of the conditions. In the context of democratic decision-making, it is obviously impossible to relax the D condition. Moreover, weakening R would defeat the purpose of this study, because it would be hard to reconcile the notion of the national interest and the unitary-actor assumption with an inconsistent preference order. Moreover, as long as one insists on employing the majority rule, nothing would be gained by modifying the P and I conditions. Consequently, it is not surprising that most approaches weaken the U condition. In fact, the assumptions of a single leader and consensus in foreign policy-making both belong to this condition, since they reduce the domain of the social choice rule to one actor or to many actors with exactly the same preferences. Under these circumstances, it becomes trivial to satisfy the other conditions, thus ensuring the existence of the national interest.

While the approach chosen in this chapter also falls under the purview of the U condition, it is less extreme than the assumption of a single leader or consensus. In this sense it closely corresponds to the traditional assumption of 'single-peaked preferences' (Black, 1948a, 1948b; Arrow, 1963). This restriction acts directly on the profile by requiring that there be an underlying strict order which does not generate any 'double peaks' in the individual orders (see Arrow, 1963, ch. 7; Kelly, 1988). Despite its analytical appeal, it seems *ad hoc* to postulate the single-peakedness property in substantive settings. Therefore, we shall start instead by imposing behaviourally plausible restrictions directly on the individual orders. The principle of 'partial unanimity' (Arrow, 1963, p. 89) requires each individual order to be compatible with a specific quasi-order, in our case motivated by substantive reasons outlined in the next section.

The constraints allow us to derive the necessary and sufficient conditions for the existence of the national interest for a specific type of conflictual situation. It should be noted that Arrow only studied sufficient conditions. Restricting their attention to the majority rule, others have introduced more-general classes of both necessary and sufficient preference restrictions (Inada, 1969; Sen, 1982, ch. 6 and 7). For our purposes, a more concrete and limited framework illustrates well how ordinal models can be reconciled with the concept of the national interest.

2 DEFINING THE INDIVIDUAL INTEREST

Before turning to the issue of aggregation, it is necessary to define individual preferences. In order to simplify the analysis as much as possible, we frame the analysis in the simplest possible way. For most states, the decision-making environment is strategic as opposed to parametric.[5] Thus, the decisions of a state are dependent on the actions of other states. Let us consider the simplest possible setting involving state i facing state j: both actors have only two strategies, to co-operate (C) or to defect (D). Figure 3.1 presents the outcome matrix which consequently contains four outcomes. Seen from player i's perspective, the outcomes are mutual co-operation (c), victory (v), defeat (d) and war (w). In a typical deterrence situation, co-operation corresponds to the status quo. Any other outcome represents some type of deterrence failure.[6]

Figure 3.1 The outcome matrix seen from the perspective of player i.

		State *j*	
		C	D
State *i*	C	c	d
	D	v	w

The precise interpretation of these outcomes is not important for the argument. Instead of deterrence, we could, for example, envisage the interaction as a negotiation situation where strategies stand for concessions versus demands, and w denotes 'no treaty' rather than war. The crucial point is that the strategies form a coercive scale from lesser to higher levels of conflict. This creates an interesting asymmetry between the outcomes that we shall exploit to constrain the universal domain condition (U).[7]

Like Arrow (1963), we assume that all individual actors are endowed with a strict preference order P_i over the four outcomes.[8] The goal is to derive a restricted individual order. Clearly, there are 4! = 24 possible orderings over these outcomes. However, all these rankings are not equally plausible from an empirical standpoint. This is where the issue of substantive rationality enters the picture. Arguing that 'thin' consistency criteria such as transitivity is not enough to define 'substantive' rationality, Jon Elster (1983, p. 15) appeals for a 'broader' notion of the concept:[9]

> Nothing has been said up to now that would prevent us from speaking of suicide, homicide or genocide as rational behaviour. Nor have I given any reason for excluding from the domain of the rational the rain dance of the Hopi, consulting the horoscope before investing in the stock market, or going back to the house rather than crossing a black cat in the street. Logically speaking, it is possible that the whole world is engaged in a conspiracy to thwart my efforts to get redress, and in the thin sense it may be

quite rational for me to act on this assumption. But clearly this sense is too thin. We need a broader theory of rationality that goes beyond the exclusively formal considerations. . .

In the foreign policy situation considered in this chapter, substantive rationality translates into the tradeoff between winning and avoiding damage. Here we exploit the assumption of coercive scales by postulating that all actors prefer winning to losing ($vP_i d$). In addition, we also postulate that, *ceteris paribus*, an actor will always want to reduce the damage. In other words, given the own-choice of strategy, the more conciliatory the strategy chosen by the opponent the better, or in formal terms: given C then $cP_i d$, and given D then $vP_i w$. We summarize these preference restrictions:

PR_1: $cP_i d$
PR_2: $vP_i w$
PR_3: $vP_i d$

where $xP_i y$ indicates that actor i strictly prefers x to y.[10] Technically speaking, these orders form a quasi-ordering which is an incomplete but transitive preference order (Arrow, 1963, ch. 4).

Although intuitively appealing, these restrictions rule out a number of potentially relevant orders. Most interpretations of pacifism, for instance, tend to violate PR_2 because defeat is preferred to the use of violence under any circumstances. To the extent that the other extreme – sadistic actors – prefer war to victory because they enjoy fighting, they would also fall outside the preference restrictions. A more common example would be the 'military–industrial complex' that might prefer war to victory in order to promote their own parochial economic interests. Such orders would contradict PR_3 as in the game of 'Bully' (see Snyder and Diesing, 1977). But for most purposes, it is safe to assume that there is at least a partial consensus concerning these tradeoffs. The restrictions should not be interpreted as a normative rejection of the rankings excluded, but rather a statement about which orders are usually empirically relevant.

Given these definitions, it becomes straightforward to filter out the orders of interest. Proposition 1 shows that only five relations satisfy the restrictions, namely Harmony (HA), Chicken (CH), Stag Hunt (SH), Prisoner's Dilemma (PD) and Deadlock (DL):[11]

Proposition 1: The preference restrictions PR_1, PR_2 and PR_3 produce five strict preference orders P_i

P_{HA}: cvdw
P_{SH}: cvwd
P_{CH}: vcdw

P_{PD}: vcwd
P_{DL}: vwcd.

PROOF: Since only strict orders are considered, the usual requirements for such a preference order, such as anti-symmetry, completeness and transitivity, are automatically satisfied. Thus, it is sufficient to enumerate all 4! = 24 permutations of the four outcomes and exclude the orders that violate one of the preference restrictions. PR_1 and PR_2 implies that the order cannot begin with d or w respectively. Twelve orders remain, namely those beginning with (1) c or (2) v.

(1) Among the former, PR_1 again constrains the possibilities by excluding orders beginning with cd. Moreover, the orders cwdv and cwvd are also disqualified because of intransitivity (cPwPdPc). The only two orders starting with c that remain are cvdw and cvwd, i.e. P_{HA} and P_{SH}.
(2) The orders starting with v cannot continue with d because of PR_3. Both orders with an initial vc are valid, i.e. vcdw and vcwd, or P_{CH} and P_{PD}. Since vwdc violates PR_1, only vwcd (P_{DL}) remains, which exhausts the possibilities.

Thus, after having ruled out 16 of the orders, we find the five orders enumerated in the proposition. QED

Figure 3.2 illustrates these five preference orders. The payoffs correspond to the ordinal ranking of each outcome, 4 being the best outcome and 1 the worst.

Figure 3.2 The five individual rational preference orders in matrix form

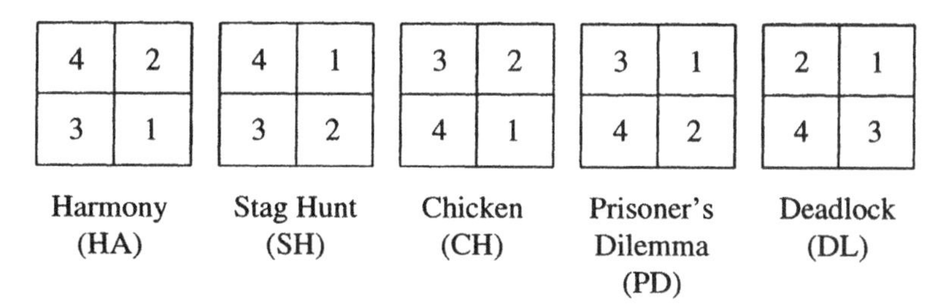

The fact that the five orders belong to the most frequently used preference configurations in the applied literature supports the usefulness of this result. In the context of deterrence, Harmony (HA) depicts an accommodating position. There is neither a first-strike incentive nor an incentive to respond in case of defeat, since war is the worst outcome. In sum, HA corresponds to a typical 'dovish' stand. Stag Hunt (SH) resembles HA but is less accommodating with respect to the defensive motives: here defeat is worse than war. Thus, we could characterize SH players as 'tough doves'. Chicken (CH) preferences represent the opposite of SH in the sense that there is a first-strike incentive. Yet losing is preferred to winning. It would not be far-fetched to label this order the 'opportunistic hawk'. It might, for instance, depict a politician in a democratic country who faces severe constraints in case of war, but who has an incentive to exploit foreign adventures for domestic purposes (Russett, 1990). The well-known Prisoner's Dilemma (PD) represents a more genuinely 'hawkish' view, with incentives both to attack and to wage war. However, this player is essentially satisfied with the status quo (c) since this outcome is preferred to war. The opposite holds for 'the expansionist hawk' whose preferences coincide with the Deadlock (DL) order.

As an illustration, consider Lebow's (1981) categorization of crises: justification of hostility, spin-off and brinkmanship. While the theory below is not dependent on both sides having the same preferences, there are important parallels between the three most conflictual preference orders and Lebow's classes of disputes. Justification of hostility corresponds closely to the DL order, since actors with this order have already made up their minds to go to war. Spin-off crises resemble PD in so far as the status quo is genuinely preferred to the war outcome. Nevertheless, because of the conflictual interests captured by the first-strike incentives, the Pareto-optimal outcome is not guaranteed. Finally, brinkmanship crises have much in common with CH, due to 'the initiator's expectation that his adversary will back down rather than fight' (Lebow, 1981, p. 57). In sum, I believe that all five preference orders correspond to theoretically meaningful and empirically relevant categories.

3 PREFERENCE AGGREGATION

Having defined the possible preference orders on the individual level, we are now ready to consider the process of preference aggregation. On the collective level, it is less important how the individual preferences were derived. While most economists favour the principle of revealed preferences, others rely on introspection and direct observation (Sen, 1982, ch. 2). In this study, it is sufficient to assume that they are given and fixed as described in the previous section.[12] Whether the preferences of the individual units have been established by inferences from the behaviour of these actors or by direct observation is immaterial.[13]

Formally, an analysis of the individual level produced a number of strict orders P_i defined over a set of social states $X = \{c,v,d,w\}$ pertaining to a conflictual situation as described in Figure 3.1. It is convenient to summarize the information contained in the individual orders as a committee profile P:

Definition 1: A committee profile P is a set of restricted individual preference orderings P_i ($i = 1, \ldots, N$) as defined in Proposition 1. Then P is completely determined by a quintuplet $(N_1, N_2, N_3, N_4, N_5)$, where the indices $N_i \geq 0$ refer to the number of individuals in Harmony, Chicken, Stag Hunt, Prisoner's Dilemma and Deadlock respectively. Furthermore, the total number of members of the committee is labelled N:

$$N = N_1 + N_2 + N_3 + N_4 + N_5 > 0.$$

In the special case $N = 1$, the committee collapses to a *unitary actor* profile.[14]

As the next step, the notion of a collective choice rule as function f is introduced (Sen, 1970). In Arrow's (1963) original terminology, such a function is called a social welfare function. Note that although the individual orders are strict (and thus denoted P), the social preferences are not necessarily strict because of the possibility of ties.

Definition 2: A collective choice rule is a function f that produces a social preference order R for a committee profile P: R $= fP$).

Since the chapter pertains particularly to democratic societies, it is natural to focus on the majority voting:

Definition 3: The majority rule is a collective choice rule f that produces a social preference order $R = f(P)$ such that:

$$\forall x, y \in X, xRy \Leftrightarrow N(xP_iy) \geq N(yP_ix),$$

where $N(xP_iy)$ stands for the number of individuals i who favour x over y.

At first sight, the choice of the majority rule may appear to be inadequate for modelling the process of foreign policy making (Achen, 1988, p. 15). Even in parliamentary systems, the national parliaments have no direct say in matters of foreign affairs. Instead, cabinets or even smaller *ad hoc* groups make the crucial decisions. In presidential systems, the president has the final say in these issues. Despite attempts to control the executive with rules such as the War Powers Act, American foreign policy is still very much in the hands of the President.

This is not to say, however, that the majority rule cannot be interpreted to capture decision-making of a less formal type than explicit voting in parliaments or other large assemblies. Suppose that the numbers N_i pertain to some power index rather than to formal votes (Snyder and Diesing, 1977, p. 349). Under this assumption, we are able to study power relationships between both individuals and organizations. In the case of the domestic American debate on arms control, we would, for instance, want to assign weights to participants in the security establishment such as the President, the National Security Advisor, the Joint Chiefs of Staff, the leaders in Congress and representatives for interest groups, and then add up the numbers of orders to form a national committee profile. An interesting possibility is to assign (sub)committee profiles to heads of organizations, the interests of which cannot be represented as one of the five individual orderings. It is well known that governmental agencies differ in terms of how they perceive the national interest (cf. Halperin, 1974, ch. 3). In such a process, the President would obviously have more 'votes' than the other actors.[15]

The crucial question of whether the national interest can be said to exist can only be answered after we have provided a clear definition of this elusive concept. Fortunately, the formal framework enables us to propose such a definition. We identify the national interest with the existence of a fully transitive preference order on the collective level. Although some theorists have suggested that this requirement is often too strong as a criterion of rationality, transitivity seems more appropriate with regard to the analysis of the national interest. An alternative would be to consider the existence of a best choice for a given agenda (Sen, 1982, ch. 6; Suzumura, 1983), but this weaker criterion would not suffice as a tool for assessing the choice quality at the national level, at least not in the long term (Maoz, 1990, p. 26). In this perspective, the applicability of the unitary rational-actor assumption hinges upon the existence of a unique, well-ordered set of preferences. Formally, we distinguish between quasi-transitive and full rational preference orders:

Definition 4: A binary relation R on a set X is a *quasi-transitive rational preference order* if

(a) it is reflexive: $\forall x \in X, xRx$;

(b) it is complete: $\forall x,y \in X, xRy \vee yRx$;

(c) its asymmetric part P is transitive: $\forall x, y, z \in X, xPy \wedge yPz \Rightarrow xPz$.

Definition 5: A binary relation R on a set X is *a full rational preference order* if it is not only a quasi-rational preference order but also transitive itself:

$$\forall x, y, z \in X, xRy \wedge yRz \Rightarrow xRz.$$

This terminology enables us to provide a formal definition of the unitary actor assumption and of the national interest:

Definition 6: Given a committee profile P and its aggregated order $R = f(P)$, the *unitary-actor assumption is* valid iff there is a full rational preference order R' such that:

$$\forall x,y \in X, xR'y \Leftrightarrow xRy.$$

Then we call R' the *national interest.*

The Condorcet voting paradox constitutes the major stumbling block of the majority rule. This is where the assumption of single-peaked preferences applies. As Duncan Black (1948a, 1948b) pointed out, and Arrow (1963, p. 77) showed formally, single-peakedness guarantees transitivity provided that the committee is of an odd size. This assumption requires the existence of an underlying strict order such that for all individuals the utility around the bliss point (e.g. the most preferred alternative) falls monotonically as one moves further away from this point. It turns out that the preference restrictions generate such a strict order, so the national interest exists for all odd-member committees.

> **Lemma**: All committee profiles P are single-peaked.
>
> PROOF: Let the underlying strict order be wvcd. Instead of using the more general formal tools suggested by Arrow (1963, p. 77) such as betweenness and so on, it is sufficient to prove that for the five individual orders from Proposition 1, the utility of alternatives to the 'left' of the best alternative increases monotonically as one moves from the left towards the bliss point. The opposite applies for the alternatives on the right side of the best alternative.
>
> *Case 1*: If the best alternative is c, then monotonicity is guaranteed iff
>
> $$\mathrm{v}P_i\mathrm{w} \wedge \mathrm{c}P_i \wedge \mathrm{c}P_i\mathrm{w} \wedge \mathrm{c}P_i\mathrm{d}.$$
>
> This is the case for the two orders that rank c as the best alternative, namely HA and SH.
>
> *Case 2*: If the best alternative is v, then we must demand that
>
> $$\mathrm{v}P_i\mathrm{w} \wedge \mathrm{c}P_i \wedge \mathrm{c}P_i\mathrm{w} \wedge \mathrm{c}P_i\mathrm{d}$$
>
> which turns out to hold for the remaining orders: CH, PD and DL.
>
> Thus, the assumption of single-peaked preferences holds for all restricted profiles. QED

Having verified single-peakedness, I turn to the necessary and sufficient conditions for applying the unitary rational-actor assumption. Proposition 2 shows that in almost all cases corresponding to the restricted preference profile, the unitary-actor assumption applies:

> *Proposition* 2: For all committee profiles P with N members, the majority rule is compatible with the unitary-actor assumption and the existence of the national interest, except for certain profiles all with an even number of members $N > 0$:
>
> *Case 1*: $(N_1, N_2, 0, 0, N_1 + N_2)$, $\forall N_1 \geq 0, N_2 > 0$

Case 2: $(N_1, 0, N_3, 0, N_1 + N_3), \forall N_1 \geq 0, N_3 > 0$
Case 3: $(N/2, 0, 0, 0, N/2)$, $\forall N = 2, 4, 6\ldots$

In these cases the social preference order is quasi-transitive rational.
PROOF: We need to prove that the social preference order R is fully rational for all restricted profiles P except for the three cases. First, we show that reflexivity and completeness hold in all contingencies. Then it is demonstrated that transitivity also applies except in exceptional cases.

Reflexivity: It is easy to prove that reflexivity holds for R. Suppose the contrary, $\sim xRx \Leftrightarrow xPx \Leftrightarrow N(xP_ix) > N(xP_ix)$, which is a contradiction, because no number can be greater than itself. Thus, R is reflexive.

Completeness: To see that R is complete, suppose the contrary, i.e. $\exists x,y \in X$: $\sim(xPy \vee yPx) \Leftrightarrow yPx \wedge xPy \Leftrightarrow \{N(yP_ix) > N(xP_iy) \wedge N(xP_iy) > N(yP_ix)\}$ $\Rightarrow N(yP_ix) > N(xP_iy) > N(yP_ix)$ which is also a contradiction. Hence, R must be complete.

Transitivity: The third criterion is less obvious since the social choice rule is majority voting. First, we observe that the preference restrictions PR_i, also hold on to the collective level: c*P*d, v*P*w and v*P*d because the restrictions apply to all individuals. To identify the cases of quasi-transitivity, we consider all candidate cycles, that is all triples. With four elements in X, there are $\binom{3}{4} = 4$ candidate (unordered) triples $(x.y,z) \in X$: (v, c, d), (v,w, d), (v, c, w) and (v, c, d). The first two never contain a cycle because of the preference restrictions PR_i. The two latter triples, which we label triple 1 and triple 2, contain a cycle iff:

Cycle 1: c*P*d*R*w*R*c
Cycle 2: c*R*v*P*w*R*c.

From PR_1 and PR_2 we know that cPd and vPw. Thus,

$$(\text{d}R\text{w} \wedge \text{w}R\text{c}) \vee (\text{c}R\text{v} \wedge \text{w}R\text{c}) \Leftrightarrow \text{w}R\text{c} \wedge (\text{d}R\text{w} \vee \text{c}R\text{v})$$

must be a sufficient condition for *intransitivity*. Yet we need to derive an expression in terms of the profiles rather than the social order. In order to do that, it is useful to divide the first term of the condition into two parts: (1) w*P*c and (2) w*I*c.

(1) The relation w*P*c is the same thing as $N(\text{w}R_i\text{c}) > N(\text{c}R_i\text{w})$ or in terms of the profile $N_5 > N/2$. But if the DL individuals are in absolute majority, the resulting social ranking must be DL which is, of course, transitive.

(2) The only remaining candidate case for intransitivity is w*I*c or $N_5 = N/2$. Again, it is necessary to check the implications of d*R*w and c*R*v. The first relation holds iff $N(\mathrm{d}P_i\mathrm{w}) \leq N(\mathrm{w}P_i\mathrm{d}) \Leftrightarrow N_2 + N_4 + N_5 \leq N/2$. Due to $N_5 = N/2$, this only applies for equality $N_2 + N_4 + N_5 = N/2$. Substituting $N_5 = N/2$ yields $N_2 + N_4 = 0$. However, since $N_i \geq 0$, we conclude that $N_2 = N_4 = 0$, which implies that $N/2 = N_1 + N_3 = N_5$. This is Case 2 in the proposition. The other possibility, c*R*v is equivalent to $N_3 + N_4 + N_5 = N/2$, is again because N_5,$= N/2$ excludes c*P*v. By similar reasoning, it is seen that $N_3 = N_4 = 0$ and $N/2 = N_1 + N_2 = N_5$, that is, Case 1. Finally, if both dRw and c*R*v, $N_2 + N_4 + N_5 = N/2$ and $N_3 + N_4 + N_5 = N/2$. In this case we have $N_2 = N_3 = N_4 = 0$ and $N_1 = N_5 = N/2$ in accordance with Case 3.

Having identified the necessary and sufficient conditions for intransitivity to be w*I*c $\wedge$ (d*I*w $\vee$ c*I*v), we realize that the candidate cycles contain one strict inequality because of the preference restrictions and the two indifferences. This means that *R* is quasi-transitive for the exceptional cases, and is always acyclic. QED

The intuition behind Proposition 2 is not as far-fetched as it may seem. Inspection of the preference matrix shows that, because of the preference restrictions, there are only two candidate cycles: one in counter-clockwise direction in the upper left corner involving d, c and w; the other one is the mirror image of the first. The cycle goes clockwise, involving c, w and v. It is immediately clear that no cycle can emerge if c*P*d which corresponds to all individual orders other than DL. Yet if d*P*c, DL is in majority, and we have w*P*d and v*P*c. These relations block the candidate cycles. The only remaining possibility is c*I*w. Now since the DL individuals control exactly half the votes, it can never be that d*P*w or c*P*v, so the only remaining possibility of a cycle is either d*I*w or c*I*v or both. These are the three exceptions of Proposition 2.

The proposition immediately yields a couple of important observations that are presented as corollaries. If the national interest is not representable by a unitary preference order, how many orders do we need? The first implication shows that in the three exceptional cases, two strict orders suffice:

COROLLARY 1: In the three exceptional cases, the unitary actor is representable as two *strict* individual orders:

Case 1: Deadlock and Stag Hunt
Case 2: Deadlock and Chicken
Case 3: Deadlock and Harmony.

PROOF: Choose the following profiles *P* for the three exceptional cases (0, 1, 0, 0, 1) (0, 0, 1, 0, 1) and (1, 0, 0, 0, 1) respectively. QED

The second result pertains to odd-member committees. It follows trivially from the definition of the single-peakedness property that a transitive preference order exists. In other words, the unitary actor assumption applies for these cases.

> COROLLARY 2: For all committee profiles P where N is odd, the majority rule generates a full rational, social preference order R which is either of the five individual orders HA, SH, CH, PD or DL.

> PROOF: The exceptional cases in Proposition 2 only occur if N is even. Thus, R is transitive. This also follows directly from the lemma about single-peaked preferences. The odd number of members rules out any draws. Then, for each of the binary choices x,y that are not restricted by the preference restrictions, either xPy or yPx. This means that all social orders must be strict. Proposition 1 shows that there are only five such strict orders, that satisfy the preference restrictions, namely: HA, SH, CH, PD and DL. QED

The exceptional types of committees for which there are no transitive order form a comparatively limited class. It makes sense that the 'schizophrenic' committees with an equal number of individuals in Harmony and Deadlock (e.g. Case 3) cannot be summarized as one individual. The only thing we can do in these cases is to reduce such committees to two individuals, one of which is in Harmony and the other one in Deadlock. Yet nothing prevents us from deriving a unique social order for groups in which there is a stalemate between Harmony and Prisoner's Dilemma (that is, profiles of type ($N/2$, 0, 0, $N/2$, 0)). In these cases the national interest is represented by the order (vc)(dw). Of course, this order may paralyse the decision-making body since it contains two indifferences, but it is not intransitive.

Corollary 2 by no means ensures that the resulting aggregated order coincides with any of the individual actors. Clearly, in the event that one of the individual orders controls the majority, the national interest will coincide with this order. In less trivial cases, however, the national interest may be different from the individual preferences. The most obvious case is a combination of individual orders producing a tie on the collective level. For instance, the national interest derived from the committee (0, $N/2$, $N/2$, 0, 0) is paralysed with respect to the tradeoff between c and v as well as d and w. Nevertheless, there is a transitive order that captures the preferences of the majority, namely (cv)(dw).

Interestingly, a similar phenomenon sometimes arises even for strict individual orders. Take, for instance, the profile ($N/3$, 0, $N/3$, 0, $N/3$), that is, is an equal mix of Harmony, Chicken and Deadlock. This committee profile, in which none of the actors enjoys absolute majority, is equivalent to a unitary actor with the Prisoner's Dilemma order, although no such individual is present in the profile. This finding is perfectly consistent with the observation that group decisions may produce unexpected outcomes: 'What happens is rarely intended and even more rarely by any one of the actor's individually' (Steinbruner, 1974, p. 141; see also Allison's Model III). That the national interest can be formally defined does not free us from the responsibility of carefully studying and justifying individual preferences (cf. Achen, 1988, p. 30).

To illustrate, I provide an example of how Proposition 2 helps us to derive the collective preference order for a specific committee profile. Consider a fictive situation in which a country is involved in a foreign policy dispute with another country. To recapitulate, the outcomes c, v, d, w are interpreted as above: co-operation, victory, defeat and war. In the parliament[16] of this country the 99 seats are distributed between the four parties in the following way: the Communist Party, which promotes a conciliatory policy corresponding to Harmony, has 20 votes; the Socialists with their 29 seats are more cautious: while being in Stag Hunt because they prefer war (w) to defeat (d), they consider a compromise solution the best of all outcomes; the 45 Conservatives, however, would like to force the other side to make concessions, rather than compromise. On the other hand, they would not mind a compromise solution since it would be superior to a costly conflict. The National Front, a right-wing extremist party, is represented by five votes. Its militant policy is equivalent to Deadlock because its adherents have no interest in negotiations.

The parliamentary situation can be summarized as the committee profile (20, 29, 0, 45, 5). If we ignore the possibility of abstentions, tactical voting and coalition-making, the majority rule requires that each decision be made with absolute majority (that is, with more than 49 votes). The second corollary suggests that it is possible to substitute the committee profile by a single, strict preference order. Moreover, we know that it has to be one of the five individual orders, HA, SH, CH, PD or DL. We need only to compare three pairs of groups: first, since the National Front is in crushing minority ($N_5 = 5 < 49.5 = N/2$), it must be that cPw, thus excluding DL from consideration. Secondly, the left-wing parties are also in a minority ($N_1 + N_2 = 49 < N/2$), from which vPc follows. Hence, only CH and PD remain as plausible candidates. Finally, the third comparison ($N_1 + N_3 < N/2$) shows that only the PD order passes the test. Thus, we have found the preference order of the equivalent unitary actor. It is Prisoner's Dilemma, the order which coincides with the policy of the Conservative Party.

4 STRATEGIC RATIONALITY

In this section, the framework is extended to the strategic level. The power of the unitary rational-actor assumption becomes obvious once strategic interaction is introduced between two compound agents. Given the absence of strategic manipulation on the individual level, the previous sections have laid the ground for an analysis of the game in two stages. In a pure bottom-up fashion, the first stage generates a well-defined 'national interest' for each state. With the few exceptions mentioned in the previous section, it is always possible to summarize any committee profile as a single preference order. The second stage addresses the strategic behaviour of such actors.

At this point, it is convenient to introduce the notion of a choice function. First we define a set of agendas $\mathcal{A}$ containing agendas that are all subsets of X Given such an agenda A consisting of choices from the set of all alternatives X, the choice function indicates which alternative from the agenda will be chosen. Besides the fact that the result cannot be the null set, there is no requirement on the choice set. For instance, it is not demanded that only one option be chosen.

> *Definition 7*: A choice function for a profile R specifies a non-empty choice set $C_R(A)$ for every non-empty agenda A.

In our specific example, we use the majority choice function:

$$C_R(A) = \{x \in A \mid \forall y \in A, N(xP_iy) \geq N(yP_ix)\}.$$

A well-known theorem (Kreps, 1988, p. 13; 1990, p. 29) says that a choice function exists if and only if the underlying preference order R is acyclic. Proposition 2 proves that the national interest is always at least quasi-rational, which is even stronger than acyclicity. In other words, it is possible to switch back and forth between the two representations.

To illustrate the strategic aspects, consider again the 2×2 simultaneous-move game. Such games are most conveniently expressed in normal form (see Figure 3.3). The players' preferences are represented as utility functions $U_i(.)$, $U_j(.)$.

Figure 3.3 The 2×2 simultaneous-move game in normal form

		Player j	
		C	D
Player i	C	$U_i(c), U_j(c)$	$U_i(d), U_j(v)$
	D	$U_i(v), U_j(d)$	$U_i(w), U_j(w)$

The Nash equilibrium is defined as an outcome from which no actor has a unilateral incentive to deviate. Formally, a Nash equilibrium is a pure strategy profile (s_i, s_j) such that

$$U_i(s_i,s_j) \geq U_i(\sim s_i,s_j) \wedge U_i(s_i,s_j) \geq U_i(s_i,\sim s_j)$$

where $\sim s_{i,} \sim s_j$ stand for the opposite strategies. In terms of the choice functions, this translates into four equilibrium conditions for each one of the four outcomes:

$$\begin{aligned}
&\text{c is Nash} \Leftrightarrow \text{c} \in C_{R_i}(\{\text{c,v}\}) \wedge \text{c} \in C_{R_j}(\{\text{c,v}\})\\
&\text{v is Nash} \Leftrightarrow \text{v} \in C_{R_i}(\{\text{c,v}\}) \wedge \text{d} \in C_{R_j}(\{\text{d,w}\})\\
&\text{d is Nash} \Leftrightarrow \text{d} \in C_{R_i}(\{\text{d,w}\}) \wedge \text{v} \in C_{R_j}(\{\text{c,v}\})\\
&\text{w is Nash} \Leftrightarrow \text{d} \in C_{R_i}(\{\text{d,w}\}) \wedge \text{d} \in C_{R_j}(\{\text{d,w}\}).
\end{aligned}$$

Note that the equilibria are labelled from player *i*'s perspective such that v means that *i* wins and *j* loses.

Since the Nash equilibrium only concerns unilateral moves out of equilibrium, it is clear that the set of admissible agendas contains only two elements: $\mathcal{A} = \{\{\text{c,v}\}, \{\text{d,w}\}\}$. This fact constrains the domain even further compared to Proposition 2. The following proposition implicitly exploits this restriction on admissible agendas to conclude that the unitary actor assumption always holds:

> *Proposition 3*: In the simultaneous-move game, both players can be treated as unitary rational actors with full rational national interest $R_{i,}$ R_j.
>
> PROOF: The restriction on admissible agendas excludes the direct choice between c and w. This means that we can rationalize the choice behaviour of DL and the quasi-orders (that is, the exceptional cases in Proposition 2) as DL, which is of course a full rational order. Thus, there are no cycles, and the unitary actor assumption applies. QED

Although the proposition guarantees transitivity, weak orders may result from the aggregation process provided that the committees are of even size. For example, if both players have the profile ($N/2$, 0, 0, $N/2$, 0) there are four Nash equilibria. In total there are nine possible weak orders for each player, yielding a total of 81 games (counting those that are each others' mirror images).

Provided that the size of the committees is odd, we conclude that there are only five strict social orders, however. In addition to this, Proposition 3 collapses the orders PD and DL into one case, so only four possibilities remain. The pure strategy equilibria (denoted from A's perspective) for all the sixteen combinations of national interest R_i and R_j appear in Figure 3.4.

Apparently, there is a unique equilibrium for twelve of the combinations. In the remaining four cases, either multiple equilibria exist or no pure strategy equilibrium. The latter contingencies occur if the players are both in CH or SH. While the corresponding symmetric games, that is, Chicken and Stag Hunt, have two Nash equilibria, the asymmetric combinations of SH and CH produce a cycle.

Figure 3.4 Nash equilibria in the simultaneous-move game for combinations of national interests

R_i \ R_j	HA	SH	CH	PD/DL
HA	c	c	d	d
SH	c	c,w	—	w
CH	v	—	d,v	d
PD/DL	v	w	v	w

These outcomes have an obvious intuitive appeal. First, acquiescent row actors in Harmony have to count on being beaten by opponents with CH, PD or DL orders. Secondly, more assertive actors in SH fight rather than surrender, thus generating no definite outcome in encounters with CH types and producing war with PD and DL. If the adversary is also in SH, there are two possible outcomes: co-operation or war. Third, actors in Chicken typically defeat HA types, and occasionally also CH opponents. However, the other possible outcome for the latter case is defeat. Because of its opportunistic approach, the CH actor is easily defeated by PD and DL. Finally, these latter actors beat both HA and CH, but have to count on war in confrontations with SH, PD and DL.

Thus, if there is a risk of encountering offensive actors (that is, CH, PD and DL), HA actors may face defeat. As an alternative, the more assertive liberal view, SH, protects against such exploitation. Yet the cost of this protection is the risk of war. The opportunistic CH is ready to exploit HA and CH in some instances, but fails to secure victory otherwise. Given that defeat follows in some of these, CH implies considerable risk of failure. Finally, both the PD and DL orders will lead to victory if the opponent is less ready to defend itself. Again, there is a tradeoff involved, since disputes with more assertive adversaries will lead infallibly to war.

It should be noted, however, that the simultaneous-move game is an inadequate model of deterrence situations (Wagner, 1983). Since deterrence is a profoundly sequential phenomenon, an extensive form game is the appropriate choice. Recursive application of the choice function would produce the subgame perfect equilibrium according to the principle of backward induction. Although the framework of this chapter could easily be extended to this type of games, I refrain from offering such an example.

Before summarizing the theoretical findings, a remark concerning collective rationality is in order. It is not obvious that the decision of a collective body would produce sophisticatedly rational outcomes. First, we have ignored strategic manipulation of the individual preference orders. Secondly, even in the absence of such strategic behaviour, we cannot be sure that a collective body would succeed in making high-quality strategic decisions. While the Nash equilibrium in the simultaneous-move game requires no forward-looking calculations, other sequential game forms typically rely on non-myopic reasoning that may strain the credulity of the pluralist interpretation of the model. In other words, empirical application necessitates a much more detailed account of foreign policy decision-making.

5 CONCLUSION

In this chapter, I have attempted to show that the applicability of rational choice stretches far beyond the unitary-actor assumption. Rather than invoking the traditional postulate of single-peaked preference profiles *a priori*, this chapter proposes a set of constraints on the individual preference orders based on intuitively plausible criteria of substantive rationality. The only orderings that survive these conditions are five well-known ordinal preference matrices: Harmony, Stag Hunt, Chicken, Prisoner's Dilemma and Deadlock. It is subsequently shown that the micro-level assumptions indeed lead to a single-peaked social preference profile. This lends more credence to this often-used but seldomly substantively justified assumption. The most important result, however, is that, with few exceptions, it is possible to apply the unitary-actor assumption and to provide a consistent definition of the national interest.

However, these conclusions do not imply a blanket endorsement of the unitary-actor assumption. By contrast, the study gives a mandate to proceed with formal methods beyond the traditional focus on the state as a 'billiard ball'. Serious attention must be given to domestic politics and sub-national interests. Even more importantly, the collective-choice model presented above rests on a number of assumptions that may prove unrealistic for many real-world situations: first, the number of players is limited to two; secondly, the set of alternatives open to each actor is also restricted to two; thirdly, the model excludes strategic manipulation of the preferences. If similar preference restrictions are imposed in a 3×3 game, for instance, it can easily be shown that intransitivity may follow. In these cases, stronger assumptions are needed to guarantee single-peakedness.

As comparative politics and international relations continue to merge, it is to be hoped that there will be more models that explicitly disaggregate the state as an actor. The reason why this trend would be welcome is the serious lack of precision afflicting the traditional bureaucratic decision-making school. In reviewing Allison's model of bureaucratic politics, Bendor and Hammond (1992, p. 319) conclude that 'Model III is so complicated that virtually no propositions can be rigorously derived from it at all'. This is where rational-choice theory can make important contributions. Zeev Maoz (1990, p. 215) outlines the assumptions of such a research programme: 'The bureaucratic perspective does not question the cognitive capacity of individual decision-makers to make complex calculations. But it does question the ability of groups to make coherent choices that are intended to maximize universally accepted national interests'. This perspective opens up an exciting agenda for future research.

NOTES

1. See Cyert and March (1963, ch. 3) for a critique of similar simplifications often used in the theory of the firm. They observe that some theories equate the preference of the firm with those of the entrepreneur, whereas other theories circumvent the difficulty by assuming a consensus on the goals of the firm.
2. In more recent work, Bueno de Mesquita has shown himself willing to deviate from the unitary rational-actor assumption (see e.g. Bueno de Mesquita and Lalman, 1992).
3. Great Britain is another alleged case of the 'single leader hypothesis'. Nevertheless, Thomas Baylis (1989, p. 58) remarks that '[p]rime ministers appear to have acted unilaterally most often in matters of foreign policy – Munich, Suez, and, under Wilson, Rhodesia are examples. Yet, foreign policy is probably also an area subject to external constraint, and while the cabinet may sometimes have to swallow an impulsive prime ministerial act, cabinet committees and other subcabinet bodies appear usually to play an important part in its formation.' With some qualification this assessment holds in a more general perspective: Baylis (1989, p. 154) concludes that 'foreign policy questions . . . are more subject to being decided "monocratically" than domestic ones – perhaps in part a consequence of the greater involvement of neocorporatist interest groups and their allied bureaucracies in the latter'.
4. However, rationality does not necessarily have to be equated with transitivity. The best way to define rationality remains a controversial issue in the social sciences. We shall return to this crucial issue below.
5. In an analogy to microeconomics, the relevant decision-making situations correspond to oligopoly theory. The exception would be very weak states totally deprived of influence on their environment (cf. price-taking firms in a perfect market) or hegemons, being able to control the international environment at their own will (cf. monopoly theory).
6. Obviously, the outcome matrix of player B is the mirror image of A's matrix. Hence, victory for A is the same as defeat for B and vice versa.
7. The structure imposed by this interpretation discards a number of the 78 games suggested in Rapoport and Guyer (1966) who do not discriminate between the two strategies.
8. The assumption of strictness circumvents logical problems associated with weak relations. Sen (1982, ch. 7) makes a less stringent assumption by excluding 'unconcerned' individuals who are indifferent between all alternatives.
9. The distinction between the thin and the broad theory of rationality corresponds closely to Max Weber's ideas about *Zweckrationalität* and *Wertrationlalität*.

10. These preference restrictions are inspired by the work of Vinod Aggarwal and Pierre Allan (1992). For other systematic derivations of preference orderings in 2 × 2 games, see Brams (1992) and Snidal (1991).
11. This notation depicts the strict preference from left to right such that *xyzw* means *x*P*y*P*z*P*w*. Ties are contained in parentheses: (*xy*)*zw* denotes *x*I*y*P*z*P*w*.
12. This obviously excludes the possibility of strategic misrepresentation (see Kelly, 1988, ch. 10).
13. It should be noted, however, that the objections against the latter alternative are weakened in the case the 'individual' orders represent organizational interests rather than single decision-makers. As opposed to individuals, organizations often have clearly-stated goals and constitutions, so there may be less of a need for introspective scrutiny in these cases.
14. Note that the unitary actor profile is not the same thing as the unitary actor assumption. Proposition 2, below, shows that this assumption may be justified in other cases than unitary actor profile. However, the unitary actor profile corresponds to the assumption of a single leader.
15. Fortunately, Proposition 2, below, does not put any limits on *N*, so we are free to choose an *N* large enough to reflect the details of the power relations in the domestic arena.
16. Of course, the choice of a parliament is an arbitrary one. We could as well have chosen a foreign policy committee or even an informal elite decision-making group such as President Kennedy's ExComm during the Cuban Missile Crisis.

REFERENCES

Achen, Christopher H. (1988), 'A State with Bureaucratic Politics is Representable as a Unitary Rational Actor', paper presented at the Convention of the American Political Science Association, Washington, D.C.

Aggarwal, Vinod K. and Allan, Pierre (1992), 'Cold War Endgames', in Pierre Allan and Kjell Goldmann, *The End of the Cold War: Evaluating Theories of International Relations*, Dordrecht: Martinus Nijhoff.

Allison, Graham T. (1969), 'Conceptual Models and the Cuban Missile Crisis', *American Political Science Review*, **63**, 689–718.

—— (1971), *Essence of Decision: Explaining the Cuban Missile Crisis*, Boston, Mass.: Little, Brown.

Arrow, Kenneth J. (1963), *Social Choice and Individual Values*, 2nd edn, New York: John Wiley.

Baylis, Thomas A. (1989), *Governing by Committee: Collegial Leadership in Advanced Societies*, Albany, N.J.: SUNY Press.

Bendor, Jonathan and Hammond, Thomas (1992), 'Rethinking Allison's Models', *American Political Science Review*, **86**, 301–22.

Black, Duncan (1948a), 'The Decision of a Committee Using a Special Majority', *Econometrica*, **16**, 245–61.

—— (1948b), 'On the Rationale of Group Decision Making', *Journal of Political Economy*, **56**, 23–4.

Brams, Steven J. (1992), 'A Generic Negotiation Game', *Journal of Theoretical Politics*, **4**, 56–66.

Bueno de Mesquita, Bruce (1981), *The War Trap*, New Haven, Conn.: Yale University Press

Bueno de Mesquita, Bruce, and Lalman, David (1990), 'Domestic Opposition and Foreign War', *American Political Science Review*, **84**, 747–65.

—— (1992), *War and Reason: Domestic and International Imperatives*, New Haven, Conn.: Yale University Press.

Cyert, R. and March, James G. (1963), *A Behavioral Theory of the Firm*, Englewood Cliffs, N.J.: Prentice-Hall.

Elster, Jon (1983), *Sour Grapes: Studies in the Subversion of Rationality*, Cambridge: Cambridge University Press.

Fearon, James D. (1991), 'Counterfactual and Hypothesis Testing in Political Science', *World Politics*, **43**, 169–95.

Ferguson, Yale H. and Mansbach, Richard W. (1991), 'Between Celebration and Despair: Constructive Suggestions for Future International Theory', *International Studies Quarterly*, **35**, 363–86.

George, Alexander L. and Keohane, Robert O. (1980), 'The Concept of National Interest: Uses and Limitations', in Alexander L. George, *Presidential Decisionmaking in Foreign Policy: The Effective Use of Information and Advice*, Boulder, Col. : Westview Press.

Halperin, Morton H. (1974), *Bureaucratic Politics and Foreign Policy*, Washington, D.C.: Brookings Institution.

Inada, K. (1969), 'On the Simple Majority Decision Rule', *Econometrica*, **37**, pp. 490–506.

Kelly, Jerry S. (1988), *Social Choice Theory: An Introduction*, Berlin: Springer Verlag.

Krasner, Stephen D. (1978), *Defending the National Interest: Raw Materials Investments and U.S. Foreign Policy*, Princeton, N.J.: Princeton University Press.

Kreps, David M. (1988), *Notes on the Theory of Choice*, Boulder, Col.: Westview Press.

—— (1990), *A Course in Microeconomic Theory*, New York: Harvester Wheatsheaf.

Lebow, Richard Ned (1981), *Between Peace and War*, Baltimore, Md: Johns Hopkins University Press.

Luce, R. Duncan and Raiffa, Howard (1957), *Games and Decisions: Introduction and Critical Survey*, New York: John Wiley.

Mann, Thomas E. (1990), 'Making Foreign Policy: President and Congress', in Thomas E. Mann, *A Question of Balance: The President, the Congress, and Foreign Policy*, Washington, D.C.: Brookings Institution.

Maoz, Zeev (1990), *National Choices and International Processes*, Cambridge: Cambridge University Press.

Rapoport, Anatol and Guyer, Melvin (1966), 'A Taxonomy of 2 × 2 Games', *General Systems*, **11**, pp. 203–14.

Richards, Diana, Morgan, T. Clifton, Wilson, Rick K., Schwebach, Valerie L. and Young, Garry D. (1992), 'Good Times, Bad Times and the Diversionary Use of Force: A Tale of Some Not-so-free Agents', paper presented at the International Studies Association, Atlanta.

Russett, Bruce (1990), *Controlling the Sword*, Cambridge, Mass.: Harvard University Press.

Sen, Amartya K. (1970), *Collective Choice and Social Welfare*, San Francisco, Cal.: Holden-Day.

—— (1982), *Choice, Welfare, and Measurement*, Oxford: Basil Blackwell.

Snidal, Duncan (1991), 'Relative Gains and the Pattern of International Cooperation', *American Political Science Review*, **85**, 701–26.

Snyder, Glenn H. and Diesing, Paul (1977), *Conflict among Nations: Bargaining, Decision Making, and System Structure in International Crises*, Princeton, N.J.: University Press.

Steinbruner, John D. (1974), *The Cybernetic Theory of Decision,* Princeton: Princeton University Press.
Suzumura, Kotaro (1983), *Rational Choice, Collective Decisions, and Social Welfare,* Cambridge: Cambridge University Press.
Wagner, R. Harrison (1983), '*Theory of Games and the Problem of International Cooperation*', American Political Science Review, **77**, 330–46.

PART II

Interdependent Preferences and Rational Choice

4. Interdependent utility functions: implications for co-operation and conflict

Michael Nicholson

INTRODUCTION

Standard analysis in the decision-theory tradition assumes that the actors are simple-minded, selfish individuals who are interested only in their own gain. In conflict models such as Prisoner's Dilemma, Chicken, and all the other models familiar in decision theory which are such a prominent feature of the genre, the actors are unconcerned about other people's utility: they get no pleasure either from other people's happiness or unhappiness. They are indifferent to other people's concerns. The same is allegedly true of groups, including the important social group, for our purposes, of the state. Simple-minded self-interest is all that matters – indeed, it is positively enjoined by those Realist theorists who slip over into moral advocacy.

Technically, altruism, or benevolence, and malice can be assumed to be built into decision-theory models. Analysts do not enquire into the origins of the utility. They merely observe (or in practice assume) that the utility is what it is. Thus the utility can include feelings of benevolence or malevolence towards the opponent. These are factored out of the problem from the beginning. However, this means that a whole range of important issues are left undiscussed where the conclusions from these issues are far from self-evident.

Hence we shall abandon the assumption that the actors are purely selfish and are interested just in some basic payoffs, where the utility is determined by these primary factors alone. We shall suppose that an actor's overall utility is determined also by the primary payoffs to the other party. These may be determined by the context of the situation. Thus, actors may develop hostility towards their rivals such that winning against the rival becomes a goal of its own, generating utility irrespective of the basic outcomes involved. Similarly, actors might develop benevolence towards an adversary and incorporate positive utilities of the rival into their own utility function. These features are part of the structure of the situation beyond the formal structure of the game itself. This can alter the structures of the games played and therefore of the strategies which it is appropriate to follow.

Though interpreted above in terms of individuals, the same sorts of issues are involved in state behaviour. In the case of states who are essentially in competition, the benefits of their rivals are seen as possibly harmful just because they are benefits, even though the direct effect on the perceiver might be non-existent. Thus, during the phase of European imperialism and expansionism which reached its peak in the nineteenth century, the dominant motive of the European powers was to get as much territory, preferably economically rich territory, under their control as possible. However, they also saw themselves as being in a race with each other where it was important for each actor not merely to get a lot of territory but to get more than the others. Some of this racing element came from the strategic issues which quickly followed on from the directly economic issues. Some also came from the desire for relative dominance in the international political economy. While mercantilist theory, preaching a zero-sum economic world, was totally discredited, hints of something close to mercantilism were to be found in the rhetoric of many statesmen when they spoke of the need to be the leading power. The imperial relations between the powers during this period involved urges to dominance. Multinational corporations seem to be by no means innocent of such motives today.

An area where the benevolent side of this analysis seems appropriate is in the possibly growing feeling that richer countries owe some obligation to poorer countries. While clearly a lot of aid, including such things as famine aid, have ulterior motives and are guided more by long-term self-interest than by altruism, it would be unduly cynical to assume that there is no altruism involved. Thus, utility functions do exist which have as one of the arguments the utility of the members of other states, even if the relevant parameters are at a regrettably low level.

The security of states *vis-à-vis* each other, particularly where they involve competitive arms races is another case where the benefits of one relate directly to the benefits (or their opposite) of the other. This raises some issues which we shall deal with towards the end of the chapter.

First, we must define the central concepts more closely. The basic payoffs of the game, say in a game of Prisoner's Dilemma, will be referred to as the 'primary payoffs'. That part of B's payoffs which generate utility, whether positive or negative for A, will be called the 'contextual payoffs'. Thus, suppose two people share a cake. The primary payoff for A is the piece of cake A actually eats and enjoys, whereas the contextual payoff is the pleasure of A in the pleasure derived by B from the part of the cake eaten by B. The problem will be discussed in terms of 'benevolence', 'malice' and 'superiority'. There may be other independent though related concepts but they will not be dealt with in this paper. 'Benevolence' is the utility an actor derives from the benefits of the partner. Clearly, this is common in many social situations. We can represent the utility function of A as $U = U(x, \mu y)$ where x is the primary payoffs of the game to A (for example, money), y is the primary payoffs to B, where μ is put in to emphasize that the contextual payoffs to A are not necessarily on a par with the primary payoffs of the partner. Hence, dU/dy is positive. The implications of this for the interpersonal comparisons of utility are dealt with below. 'Malice' (or 'malevolence') is the negative of benevolence. It is when one person is upset by their partner's good fortune (a common enough feature of life). It can be represented by the same general form of function, only in this case dU/dy is negative. 'Superiority' is the desire to win in itself, so a utility function is of the form $U(x, \{x - y\})$. 'Benevolence' and 'malice' are clearly symmetrically related whereas 'superiority' is something different. Two other forms of interdependent utility functions are suggested by considerations of symmetry with 'superiority', namely $U(x,\{y - x\})$ which represents 'inferiority'. This might be the case for a very humble person or for someone in a hierarchic society who 'knows his place'. Another would be $U(x, \{x + y\})$ where actors gain utility from the aggregate and also from their own share of it. However, in this chapter I shall not consider these last two forms but merely draw attention to them.

This general issue has interested a number of international-relations scholars recently (Grieco, 1991; Nicholson, 1990 and 1992; Powell, 1991; Snidal, 1991a and 1991b). Snidal's first 1991 paper (1991a) is of particular relevance to this chapter and in a slightly different form we shall discuss some of his results and also relate them to a wider form of analysis. The issue has also been discussed in a rather different context by Taylor (1987). Taylor is concerned about co-operation under anarchy, though with little specific reference to the implications of his work to international relations.

1 SUPERIORITY AND CHANGING GAME STRUCTURES

Let us suppose we have a game which has basic payoffs ranked by the actors according to the Assurance game,[1] which has the following matrix:

Matrix 1

		II: C	II: D
I	C	(4,4)	(1,3)
I	D	(3,1)	(2,2)

These rankings are those of the primary payoffs. It is convenient to think of them simply as money outcomes. The central part of the analysis will require them to be cardinal utilities.

If we use the convention that the earlier a letter appears in the alphabet the higher up the scale of preferences the outcome is for the particular actor, we can represent the game more generally as:

Matrix 2

		II C	II D
I	C	(a_1, a_2)	(d_1, b_2)
I	D	(b_1, d_2)	(c_1, c_2)

However, if the actors also feel hostility towards each other, then gaining a higher value than the rival is something they also value. In the limiting case, all they would be interested in would be gaining an advantage, so that to be superior to the rival overwhelms all other considerations, that is, the values in the basic payoff matrix. In such a case, the payoff matrix would then consist of the differences between the basic payoff of the player and the rival's payoff. Matrix 1 will become matrix 3:

Matrix 3

		II C	II D
I	C	(0,0)	(–2,+2)
I	D	(+2,–2)	(0,0)

By definition this is a zero-sum game, in which both players use the defect strategy, with the result or solution of (0, 0). The interpretation of the names of the strategies 'C' and 'D' as 'co-operate' and 'defect' begins to look rather odd, but it is convenient to keep the letters as names.

This is an extreme assumption. It assumes that the only goal the actors have is to be superior to the opponent in this particular sense. A common situation is more likely to be between these two extremes where the primary value of an outcome, plus the advantage achieved over the rival, are both relevant. We can represent these for A alone in the general matrix as follows. In this, λ is the 'superiority factor' for A:

Matrix 4

		II	
		C	D
I	C	$(a_1 + \lambda[a_1 - a_2])$	$(d_1 + \lambda[d_1 - b_2])$
	D	$(b_1 + \lambda[b_1 - d_2])$	$(c_1 + \lambda[c_1 - c_2])$

We can show that if λ is sufficiently large the game changes from Assurance to Prisoner's Dilemma, while if λ is very large so that the difference term totally dominates the basic outcome, the game approaches a zero-sum game. This can be illustrated in terms of the basic Assurance game illustrated in matrix 1 and adding in a factor for superiority. We assume the game is symmetrical not only in basic payoffs but also in degree of the desire for superiority. This assumption is only for convenience and is not crucial for the analysis.

Matrix 5

		II	
		C	D
I	C	$(4 + 0\lambda, 4 - 0\lambda)$	$(1 - 2\lambda, 3 + 2\lambda)$
	D	$(3 + 2\lambda, 1 - 2\lambda)$	$(2 + 0\lambda, 2 - 0\lambda)$

If λ is zero, then this is the basic Assurance game from which we started. However, suppose $\lambda = 1$. In this case we have the payoff matrix of:

Matrix 6

		II	
		C	D
I	C	(4,4)	(–1,5)
	D	(5,–1)	(2,2)

This is a Prisoner's Dilemma game.

If we go yet further and assume that $\lambda = 100$, then the matrix becomes:

Matrix 7

		II	
		C	D
I	C	(4,4)	(–199,203)
	D	(203,–199)	(2,2)

This is clearly approaching zero sum and gets closer to it the greater we make λ. The ratios of I's payoff to II's payoff when CD or DC are played get closer and closer to –1 as λ increases, while the entries in CC and DD become trivial in comparison. Strictly, the game always remains a Prisoner's Dilemma, except in the limit as $\lambda \to \infty$, though the payoffs involved are rather extreme.

From this it is clear that by some appropriate choice of a value of λ, we can transform the game, which in its primary payoffs is the highly co-operative Assurance game, into other games of varying degrees of competitiveness right up to the totally competitive zero-sum game. To know which game we are playing we need to know the value of the λ parameter, which I have called the dominance parameter. It is interesting to note, however, that there is no choice of λ possible which will transform the Assurance game into a Chicken game. It moves directly from Assurance to Prisoner's Dilemma.

BENEVOLENCE AND MALICE

Snidal (1991a) has discussed the situations which are here called malice in some detail. I shall just give a general outline of the issues for comparison. Clearly, the underlying results are the same, but Snidal's presentation of the problem entirely in terms of malice understates some of the issues we are concerned with.

A utility function involving malice can be represented as $U = x - \mu y$, which is the form Snidal discusses. We shall also deal with the direct obverse of malice which is benevolence, where $U = x + \mu y$. The most interesting result for our purposes is that as μ increases over the interval (0,1), all types of games, whether Chicken, Assurance, Harmony or whatever, turn into Prisoner's Dilemma. Though $\mu = 1$ is severe malice there is no analytical reason why this should operate as a restriction, and the result does not hold for μ sufficiently large, as I show by an example. In particular, $\mu > 1$ might very plausibly hold in cases where x was considerably larger than y.

First let us consider a Prisoner's Dilemma with malice starting off with the 'basic' Prisoner's Dilemma:

Matrix 8

		II C	II D
I	C	(3,3)	(1,4)
I	D	(4,1)	(2,2)

Suppose first that $\mu = \frac{1}{2}$, The resulting derived matrix is:

Matrix 9

		II C	II D
I	C	$(1\frac{1}{2},1\frac{1}{2})$	$(-1,3\frac{1}{2})$
I	D	$(3\frac{1}{2},-1)$	(1,1)

This is still Prisoner's Dilemma. If malice gets greater and $\mu = 1$, then we get a zero-sum game of straight differences which can also be looked at as a form of degenerate Prisoner's Dilemma, namely:

Matrix 10

		II C	II D
I	C	(0,0)	(–3,3)
I	D	(3,–3)	(0,0)

Clearly, the equilibrium point of this, in either a single shot or a supergame, is the bottom right-hand box (the DD strategy).

Now consider malice in the case of the Assurance game. Again, we assume the linear utility function, though now setting $\mu = \frac{1}{2}$. Consider the following Assurance game:

Matrix 11

		II C	II D
I	C	(8,8)	(2,6)
I	D	(6,2)	(4,4)

The derived game from this matrix can quickly be seen to be identical with the Prisoner's Dilemma game of matrix 6. Thus we have turned the Assurance game into a Prisoner's Dilemma. This is what we would expect from Snidal's analysis.

However, if we relax Snidal's conditions and allow $\mu = 2$, the matrix is transformed into:

Matrix 12

		II C	II D
I	C	(–8,–8)	(–10,2)
I	D	(2,–10)	(–4,–4)

In this case the DD strategies are dominant and, more interestingly, point DD is Pareto optimal.

Now let us consider a case where benevolence is operative. We shall take the case of a simple linear function, namely that the utility to A is $x + \mu y$ and the utility to B is $y + \mu x$. Consider the following example of Prisoner's Dilemma:

Matrix 13

		II	
		C	D
I	C	(4,4)	(1,6)
	D	(6,1)	(2,2)

If we suppose that $\mu = \frac{1}{2}$ the derived payoff matrix will be:

Matrix 14

		II	
		C	D
I	C	(6,6)	$(4,6\frac{1}{2})$
	D	$(6\frac{1}{2},4)$	(3,3)

This, rather surprisingly, is Chicken.

However, suppose that matrix 13 is replaced by another Prisoner's Dilemma, namely:

Matrix 15

		II	
		C	D
I	C	(4,4)	(1,5)
	D	(5,1)	(3,3)

In this case the derived matrix for $\mu = \frac{1}{2}$ is:

Matrix 16

		II	
		C	D
I	C	(6,6)	$(3\frac{1}{2},5\frac{1}{2})$
	D	$(5\frac{1}{2},3\frac{1}{2})$	$(4\frac{1}{2},4\frac{1}{2})$

This is Assurance.

If we had an extremely benevolent case where people regarded their partner's utility as equivalent to their own, meaning that μ is equal to 1, then the derived matrix from matrix 13 would have been:

Matrix 17

		II	
		C	D
I	C	(8,8)	(7,7)
	D	(7,7)	(4,4)

The CC point in this game is very stable. This game is where the actors maximize their joint payoff. It results when side-payments are permitted and either both players trust their colleague to honour side-payment agreements or there is some enforcement mechanism.

Thus, we now have situations where, as μ moves from zero to one, we switch from a Prisoner's Dilemma game to a Chicken game to a game with a very strong collaborative outcome which would be hard to fault. This contrasts with the superiority case.

3 RELATIONS BETWEEN THE DIFFERENT FORMS OF UTILITY FUNCTIONS

A number of different forms of interdependent utility functions are plausible for the purposes of this general analysis. We need to consider the relationships between them in greater detail.

First, let us now consider the relationship between superiority and malice in the case of the linear functions employed above. In the case of superiority we have the utility function of the form $U = x + \lambda(x - y)$ with $\lambda > 0$. For malice we have $U = x - \mu y$ where $\mu > 0$, and benevolence where $\mu < 0$. Let us consider the relationship between superiority and malice by setting them equal, that is:

$$x + \lambda(x - y) = x - \mu y. \tag{4.1}$$

From this it clearly follows that:

$$\lambda = -\mu\, y/(x - y). \tag{4.2}$$

Thus, except for the small and not particularly interesting set of games where $y/(x - y)$ is constant, games involving superiority are not equivalent to games of malice (and therefore benevolence). The relationship appears even more clearly if we express malice in a slightly different form and represent the utility function as a linear combination of the primary and the contextual payoffs. Let $U = (1 - \mu)x + \mu(x - y)$. Clearly, this can be rewritten as $x - \mu y$ and is hence the same thing as malice. This is the form of the utility function that both Taylor and Snidal use. The analysis of both superiority and malice is similar and some of the results are related. The phenomena are different, however.

Figure 4.1 (a)

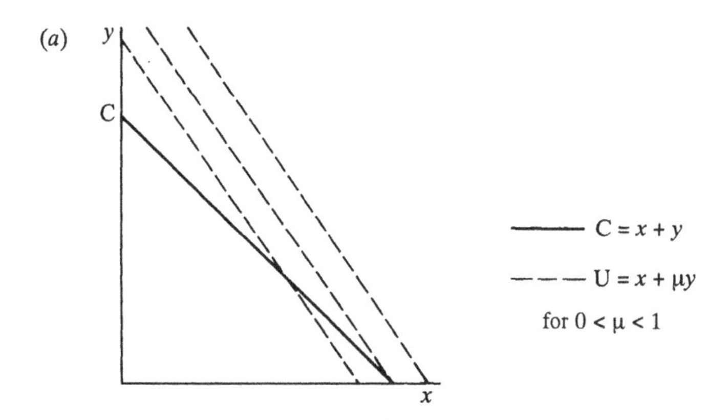

The linear utility function makes the analysis simple but it has some weaknesses both at the intuitive level and at the level of giving counterintuitive results. $U = x + \mu y$ in effect gives a set of linear indifference curves between x and y as illustrated in Figure 4.1(a). The obvious weakness with this is that it gives the same rate of exchange (that is, marginal rate of substitution) between x and y irrespective of their various relative proportions. This does not seem very plausible. It further gives counterintuitive results in some cases.

Figure 4.1 (b)

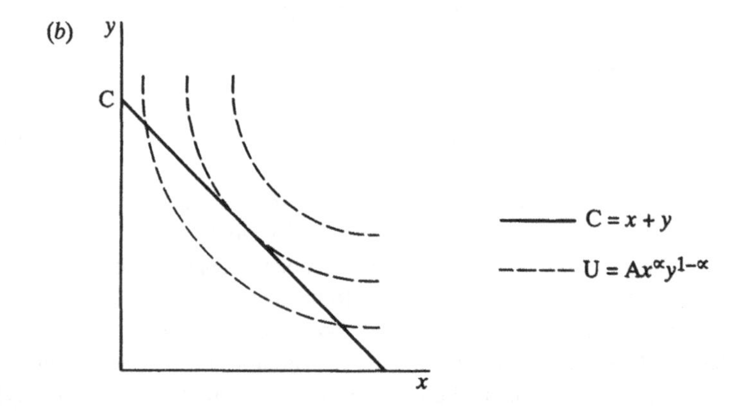

Suppose we have a constant game of division (of, say, a cake) where the division is made by actor A (whose utility is x), who has some degree of benevolence given by the function $x + \mu y$ ($\mu > 0$). The various possible divisions of the cake between A and B are given by $C = x + y$, where C is the size of the cake. Clearly, there will be a corner solution; which corner it is depends on whether μ is greater or less than unity (or there may be an indeterminate solution if $\mu = 1$). In either case the solution seems to miss out some of the more plausible outcomes.

The problem is surmounted if we adopt the hypothesis that the indifference curves between the two utilities are convex to the origin, as normally assumed about indifference curves when they relate conventional variables. The assumption seems equally plausible in this context and with this interpretation of the variables. A function widely used in this sort of analysis is

$$U = Ax^{\alpha} y^{1-\alpha} \qquad \text{where } 0 < \alpha < 1. \tag{4.3}$$

This gives a set of indifference curves of the conventional shape. The closer α is to unity the more it values its own utility and the less it values its companions, and vice versa with the limiting case where $\alpha = 0$.

How does this affect the earlier analysis? Consider the relationship between μ and α in the two different utility functions. First set

$$x - \mu y = A\, x^{\alpha} y^{1-\alpha} \tag{4.4}$$

From this we can rearrange the terms and see that:

$$\mu = x/y - A(x/y)^{\alpha} \tag{4.5}$$

From this we see that we have $\mu > 0$ (and hence malice) if

$$(x/y)^{1-\alpha} > A \tag{4.6}$$

while we have $\mu < 0$ (and hence benevolence) if

$$(x/y)^{1-\alpha} < A \tag{4.7}$$

Instead of being a constant, as in the treatments of the problems so far, μ is now a variable which is a function of (x/y). As in any conflict situation the relative values of x and y are likely to change (indeed, only unusually would they not). It is quite possible for the utility functions to exhibit malice at some points and benevolence at others. We can demonstrate this easily by the simple case where $A = 1$. Here equation (4.4) becomes:

$$x - \mu y = x^{\alpha} y^{1-\alpha} \tag{4.8}$$

and equation (4.5) becomes

$$\mu = (x/y) - (x/y)^{\alpha}. \tag{4.9}$$

From this we can see that $\mu > 0$ (malice) if $x > y$, and $\mu < 0$ (benevolence) if $x < y$. This is illustrated in Figure 4.2.

Figure 4.2

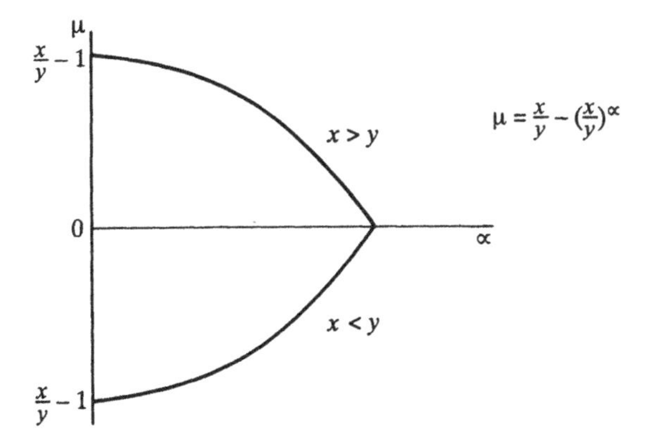

Malice and benevolence are not very appropriate terms in this context; such a utility function would in fact imply a profound sense of fairness. If A has more than B (that is, $x > y$), A positively values B's utility and the bigger the advantage becomes the more the colleague's utility is considered. Conversely if A has less than B, A negatively values B's utility.

4 THE MEANING OF INTERDEPENDENT UTILITY FUNCTIONS

Consider the interpretation of the expression $U = x + \mu y$, initially supposing that μ is positive, but we shall go on to examine various cases.

First, consider the utility function as applied to an individual. Suppose we interpret the utility function as implying wellbeing, happiness or whatever, so that the general interpretation is that there is some element of a 'self-regarding' utility, that is the utility due to the pleasure experienced directly by the actor, and another element formed by the pleasure in someone else's experiencing pleasure. It is a common experience to derive pleasure from other people's pleasure. Indeed, it is difficult to see what personal affection or love would mean if it did not involve pleasure in someone else's pleasure. Thus, in dividing the cake we spoke of earlier between two people, the standard actor in decision theory is supposed to want the largest slice possible: the ideal solution would be to have the entire cake. However, in many social situations one actually derives pleasure from the other person's pleasure and prefers that they should have some, even if one enjoys the cake very much for its own sake. Other considerations may come into play. Even though one may not particularly like one's companion, one may still be moved by such factors as a sense of justice and fairness. Carrying out a just act is something which also gives utility to actors, and thus they may prefer their companions to have some of the cake rather than always being self-regarding egoists. It might also involve altruism where, though one gets no direct pleasure from someone else's pleasure, one gets utility from the feeling of having acted virtuously, from having contributed to a more just society and so on. Conversely, of course, we may dislike the person and gain positive pleasure out of depriving them of the cake. Such factors seem so obviously to exist in personal behaviour that the analysis of a lot of behaviour without acknowledging it seems odd.

However, we need to examine the appropriateness of expressions such as $U = x + \mu y$ as interpretations of such phenomena. We have to be even more careful to establish the legitimacy of extending it beyond the confines of individual decision-making.

First, this 'utility' is not the same as that defined by Von Neumann and Morgenstern, but harks back to an earlier tradition of utility in economics. The

Von Neumann and Morgenstern utility is a strictly behavioural definition of how people behave under situations of risk. It may be that utility as above and that of Von Neumamn and Morgenstern behaviourally coincide. For practical purposes we shall assume that this is so and make no further distinctions; but a caveat was necessary to salve the author's philosophical conscience.

Secondly, this analysis, in which we add and subtract one actor's utility to and from another's, appears to imply the interpersonal comparison of utilities, which is always looked upon with deep and proper distrust. Let us consider it more carefully in the case of malevolence, where an argument in A's utility function is $x - \mu y$. Now the y in this case is A's *perception* of B's utility as some proportion of A's own. It is nothing to do with what it 'really' is (which here has no meaning). We are talking about the psychological state of A; that is, it is A's perception, which is a perfectly meaningful concept. Furthermore, both introspection and casual observation suggest we do it all the time. The greater weakness of the analysis is on the practical not the conceptual level. If we consider A's perception of B's utility and assume A has defined one of B's utility intervals, say $(a_2 - b_2)$, then the values of c_2 and d_2 are determined, at least in principle. Thus, and again in principle but rather less plausibly in practice, they are public knowledge. However, we are assuming that they are known to A. That is, it is not extravagant to assume that A has a perception of B's utility function; indeed, we would suppose so, but we have supposed that A's perception is correct. This is a strong assumption about the degree of knowledge available, though no stronger than is made in lots of analyses of this sort. However, the model is structurally stable so, providing the perceptions are roughly right, the model will also be roughly right.

Thirdly, and crucially, this interpretation of utility involves the psychological state of the actor. We are talking of mental states. It is perfectly legitimate as far as the applications made by Taylor are concerned, but it needs looking at more carefully when we start talking of groups. In its application to international relations, we are almost invariably discussing groups, whether governments or other actors in the international system. Clearly, in the case of a group, we cannot be referring to mental states. In the case of the Von Neumann–Morgenstern utilities there is no real problem: it is a behavioural measure. To make it meaningful here, we have to interpret the concept again, such that there is a behavioural correlate. Let us consider the problem in the context of security, which is commonly supposed to be a major goal of some of the actors in international relations. For an individual this could be interpreted as a state of mind. For a group it can be interpreted as the goals which can be inferred from the set of decisions which are made where security matters are considered to be the issue. For example, we can think of a set of indifference curves involving, say, security issues, social expenditure and disposable income. Such a set is in principle a behavioural concept. It does not matter that the decisions are made by groups.

Clearly, it is not only the preferences of the individual members of the group which determine the group's decision. Thus, from one set of individual preferences, a majority voting system within the decision-making group such as a cabinet would lead to one set of choices, whereas a two-thirds majority system would yield another. Different systems would also make a difference. A hierarchical system with a president acting, but under advice, would yield one solution, whereas normally a plebiscite would yield another. For the analysis all that matters is that choices between alternatives are made and can be coherently represented. An indifference-curve analysis merely leads to an ordinal utility. Cardinality, which is required in this analysis, can be introduced if we argue by the same procedures that the group can have perceptions about its own and other states. The simplest assumption is to assume that the utility defined as security obeys the Von Neumann–Morgenstern postulates about utility. They are robust postulates, so approximate agreement is sufficient for the analysis.

Let us now turn to the more direct issues of interpretation. Some were mentioned in the introduction but we look particularly at the issue of security, where the relative gains concept would initially appear to have most direct relevance. The first problem is fairly straightforward. If we are discussing armaments, for example, we can argue that a state's security is determined by its own armaments and those of other countries. Thus we have a 'security function' $S = f(x\ y)$ which can be analysed exactly as above. If A perceives B as an enemy, then, $\partial S/\partial y < 0$, and if an ally, then $\partial S/\partial y > 0$. More generally, for n states $S = (x_1\ x_2 \ldots x_n)$ where x_1 is its own level of arms and the other x_i $i = 2 \ldots n$ are the levels of the other states, $\partial S/\partial x_i < 0$ will be the case for perceived enemies and $\partial S/\partial x_i > 0$ will be the case for perceived allies. In effect, the first system is a Richardson model and the second is the rather more general Richardson model as analysed by Richardson (1960) himself and Schrodt (1981). It would be eccentric to deny that the levels of the arms of other actors were relevant.

However, if we take the case of security itself it is not clear what 'relative gains' would mean. Security itself in the context of a competitive grouping, such as in the various versions of the Realist theory, is essentially a relational concept. There is no such thing as security in the abstract but only in the context of the system of other actors.[2] Thus it is unlike profits in the case of an economic market, where the concept is essentially the same in perfect competition or a monopoly; security is meaningful only in relation to other actors, as is clear from the brief discussion above about armaments. If S is the security of A and T the security of B, then it is not clear what defining some super-security, say a utility U as $U = S - \mu T$, would mean because the factors determining T and S are already taken into account in their initial definitions. At best it would mean some rather arbitrary splitting up of some of the elements in security, some to go in with S and others with μT. To determine S we have already had to take into account such issues as the level of armaments of both A and B and possibly other things such as their underlying economic power, level of technical development, trading position in connection with armaments and so on. Further, in such discussions it is implied that security, if not necessarily a zero-sum relationship, is necessarily competitive, so an increase in one actor's security involves the reduction of someone else's. Clearly, there is no particular reason why this should be: the purely defensive security of one actor can increase such that its partner's also increases or at least remains constant.

5 CHANGES IN GAMES THROUGH TIME

Though the analysis we have carried out applies to single-shot games, it is more naturally interpreted in the case of repeated games where the basic game is iterated sequentially through time. In such situations, we might expect the various parameters to alter as the conflict progresses. Consider a game of the Prisoner's Dilemma in the basic matrix. Initially there might be no benevolence but they may still play the CC strategy (for example, because they had read Axelrod). However, repeated interaction might induce some feeling of benevolence and the structure of the game would alter and become a Chicken game. Repeated playing of Chicken in a cooperative way might increase yet further the feeling of benevolence (that is, μ increases) and get to the very stable state depicted by matrix 10. However, if in the Chicken stage one of the parties were tempted to defect, this would be expected to alter the structure of the game. That is, the defection would produce (we suppose) a sharp fall in benevolence on the part of B, for whom it would become a Prisoner's Dilemma again. If A had played the defect strategy, B would have responded by defecting. If players are aware of these features of the situation, they play a sort of double game. They must work out the consequences of any strategy not only in the game as it is at the moment, but also in what it might turn into as a consequence of the strategy played. The payoffs in future games are dependent, at least in part, on the previous pattern of play. Such forms of reasoning again seem to match up with intuitive ideas of how we behave in conflict situations. Thus, we might desist from a certain profitable strategy because it would anger a rival and make future co-operation harder.

Suppose, in the initial basic game of Prisoner's Dilemma, A had played the defect strategy. This would have induced malice (that is, $\mu < 0$) and created a more severe Prisoner's Dilemma game, as illustrated in matrix 8. This could (though it need not) induce further defect playing and further feelings of malice (μ becomes a larger negative number) and we might end up with the more extreme form of quasi-Prisoner's Dilemma illustrated in matrix 9, with the DD outcome as the solution point.

6 CONCLUSION

The type of problem considered in this paper has attracted interest in recent years in the international relations literature. Earlier it had been discussed in the hypergame literature (Bennett, 1977; Bennett and Dando, 1979), though the ways in which the structure of the games altered were not noted, to my knowledge, until Taylor drew attention to them. It is perhaps surprising that this feature has not received greater attention in the literature until recently. It gives some indication about how some classes of games which are nearly zero-sum can be approached. The preferences involved can be split into two constituent parts: the preferences about 'inherent' factors, that is, the primary preferences; and the preferences due to the context and attitudes towards the opponent. For many purposes, the contextual elements can be appropriately referred to as conflict-dependent, because their values are dependent on the attitudes to the rival generated in a conflict and are likely to vary with the nature of the conflict in which the parties are involved. In order to alter a game to a non-zero-sum game in which there are mutually advantageous agreements, the emotional hostility elements must first be reduced, after which the primary or substantive preferences underlying them can be approached. This does not apply to all zero-sum games, of course. Many players have preferences which are simply opposed. However, by bringing in issues of attitudes, we can show how they affect the underlying structure of the game and make them more (or less) amenable to solution. It gives us some clues as to how we might achieve some forms of conflict resolution. If we hypothesize that primary preferences are more stable than the conflict-dependent preferences then, over the course of a conflict, these will change. The role of conflict resolution is to manipulate them such that they decrease. This is the object of 'cooling-off periods' in the resolution of industrial disputes. One form of conflict resolution in international affairs is to convert malice parameters into benevolence parameters, but I would surmise that this is a longer-term process, though one which does happen, even in international relations, for example, between former 'traditional enemies' such as France and Germany. A less happy manifestation of the failure of these processes is taking place in the former Yugoslavia at present (1994).

While the dominance, benevolence and malice factors complicate this sort of game approach from a purely technical point of view, this in no way invalidates the basic analysis. The essence of the rational models, at least in the present stage of development, is to highlight the conceptual issues of conflict. An important role, for example, is to classify the different conflict types. That is unaltered by this analysis. What it does show is that the conflict types can be transformed from one sort into another with variations in the hostility level; this of course makes the analysis more complicated in that there are now more variables. It has nevertheless added something to the problem in raising another set of issues which are clearly relevant to it, and doing so in a reasonably precise form. This is one of the roles of theory – to formulate questions, and to formulate them as unambiguously as possible. It is better still if we get to the stage where we can answer the questions, that is, when we get to theory proper as distinct from the models which we are analysing at the moment; but that remains in the future.

NOTES

1. The definition of an 'Assurance game' differs amongst different scholars. I follow Sen's (1982) usage which I also adopt in Nicholson (1992). Jervis (1988) defines such games as 'Stag Hunts' and Assurance is defined as C/C (4, 4), C/D (2, 4), D/C (4, 2) D/D (3, 3). Waltz's (1979) discussion of the Stag Hunt arguably refers to a situation which I refer to as Assurance, but in Waltz the situation is insufficiently closely specified to be sure what the entries in the matrices would be. Taylor's (1987) definition of Assurance is the same as Jervis's but he calls the Sen version of Assurance a variant of Assurance. This confusion is unfortunate, though it does not affect the analysis as such.
2. There is obviously also the question of internal security, but I shall ignore it in this discussion. Related problems are discussed in Schmidt (1991).

REFERENCES

Axelrod, Robert (1984), *The Evolution of Cooperation*, New York: Basic Books.
Bennett, P.G. (1977), 'Toward a Theory of Hypergames', *OMEGA*, **5**, 749–51.
—— and Dando M. (1979), 'Complex Hypergame Analysis: a Hypergame Perspective of the Fall of France', *Journal of the Operational Research Society*, **30**(1), 23–32.
—— and Huxham, C.S. (1982), 'Hypergames and What They Do', *Journal of the Operational Research Society*, **33**, 41–50.
Grieco, Joseph (1990), *Cooperation among Nations: Europe, America and Non-tariff Barriers to Trade*, Ithaca, N.Y.: Cornell University Press.
Jervis, Robert (1988), 'Realism, Game Theory and Cooperation', *International Organisation*, **40**, 317–49.
Nicholson, Michael (1990), 'Rational, A-rational and Irrational Behaviour', paper given at the 31st Annual Convention of the International Studies Association, Washington, D.C.
—— (1992), *Rationality and the Analysis of International Conflict,* Cambridge: Cambridge University Press.
Powell, Robert (1991), 'Absolute and Relative Gains in International Relations Theory', *American Political Science Review*, **85**(4), 1303–20.
Richardson, Lewis Fry (1960), *Arms and Insecurity*, Pittsburgh, Pa: Boxwood Press.
Schmidt, Christian (1991), *Penser la guerre, penser l'économie*, Paris: Odile Jacob.
Schrodt, Philip (1981), *Preserving Arms Distributions in a Multi-polar World*, Monograph Series in World Affairs, Denver, Col.: University of Denver.
Sen A.K. (1982), *Choice Welfare and Measurement*, Oxford: Basil Blackwell.
Shubik, Martin (1971), 'Games of Status', *Behavioral Science*, **16**, 117–29.

Snidal, Duncan (1991a), 'Relative Gains and the Pattern of International Cooperation', *American Political Science Review*, **85**(3), 701–26.

—— (1991b), 'International Cooperation and Relative Gains Maximisers', *International Studies Quarterly*, **35**(4), 387–402.

Taylor, Michael (1987), *The Possibility of Cooperation*, Cambridge: Cambridge University Press

Waltz, Kenneth (1979), *Theory of International Politics*, Reading, Mass.: Addison-Wesley.

5. Preferences, beliefs, knowledge and crisis in the international decision-making process: a theoretical approach through qualitative games

Christian Schmidt

INTRODUCTION

The purpose of this chapter is to clarify the notion of 'crisis', when applied to international situations, through the use of analytical tools, mainly provided by the study of Nash equilibria in non-co-operative 2 × 2 games. Indeed, the typology of 2 × 2 non-co-operative games has already been extensively studied by game theorists over a long period (Rapoport, Guyer and Gordon, 1966; Guyer and Hamburger, 1969) and its relevance in the field of international relations has been carefully scrutinized (Brams, 1977; Brams and Wittman, 1981; Brams and Hessel, 1982; Brams and Hessel, 1984). However, the study of Nash equilibria in non-co-operative games has recently become the object of further elaboration by means of new refinements related to the meaning of assumptions on information, knowledge and cognitive structure. Such a renewal in the study of non-co-operative games emerges, first of all, from the works of economists involved in game theory (Selten, 1975, 1978; Rubinstein, 1982; Kohlberg and Mertens, 1986; Binmore and Dasgupta, 1986; Harsanyi and Selten, 1988; Binmore, 1987, 1990; Kreps, 1990; Osborne and Rubinstein, 1990). The classification of 2 × 2 games was then revised in the light of this new trend (Walliser, 1988). We therefore propose to apply this new material in order to specify the understanding of the decision-making processes in so-called 'international crises'.

The point of view adopted here is restricted to a class of phenomena in international affairs which are the result of different kinds of spurious effects in rational decision-making processes.[1] A more precise definition will be provided further on. According to this approach, three main sources can be identified in the first place:

(a) a lack of discrimination among available strategies on the grounds of rational choice;
(b) an ambiguity in the framing of the situation;
(c) a misinterpretation of the real game by the players.

Secondly, various suggestions are made for withdrawing from such crisis situations through an in-depth analysis of conditions of information, belief and knowledge.

Thirdly, a few empirical situations are sketched out as simple illustrations, though a systematic empirical investigation of the proposed framework will comprise one of the next steps in the research programme.

1 TOWARDS A PROTOTYPE OF CRISIS IN A DECISION-MAKING PROCESS

The line followed in order to explore the logical foundations of an international crisis is that of finding the simplest prototype in the case of a 2×2 model and exploring systematically its various features. We propose to start with a situation summarized by the following matrix, where I and II are two different countries and a_1,b_1 and a_2,b_2 two alternative actions (see matrix 1).

Matrix 1

		II	
		a_2	b_2
I	a_1	0,0	1,0
	b_1	0,1	1,1

Three reasons can be given for choosing matrix 1 as a prototype:

(a) matrix 1 covers all types of combination of interest between country I and country II;
(b) many interpretations of matrix 1 can be found in the field of international relations. We should mention two of them:
    (i) matrix 1 can illustrate the case of two nuclear powers in a situation of potential adversaries, for example the United States and the Soviet

Union during the Cold War period. In this case (a_1,a_2) means 'attacks' and (b_1,b_2) 'does not attack';[2]

(ii) matrix 1 is also relevant for describing a co-operation dilemma applied in various cases (Axelrod, 1984), where a_1 and a_2 now mean 'defeat' and b_1 and b_2 'co-operate' (Schmidt, 1991a).

(c) from matrix 1, almost all sources of crisis previously mentioned are tractable, as we shall now see.

Attention should be drawn to the definition of the payoffs in Matrix 1. 1 and 0 must be understood not as numbers but as indices. A possible interpretation would be, for example, 'satisfactory' for 1 and 'not satisfactory' for 0. Later, this approach will be extended to an ordinal ranking, corresponding to players' preferences; but in any case, the proposed prototype is free from any specific numerical values. We wish to point out that such a presentation is consistent with the classical 'maximization principle' as well as with the 'satisficing principle' related to Simon's conception of rationality. More precisely, the two countries considered can act as 'maximizers', or as 'satisficers'. Thanks to such features, Matrix 1 has a broad generality, even if the vector's arrangement appears arbitrary.

Finally the prototype can easily be extended to more than two countries. Let us suppose, for instance, three countries: I, II and III. A non-co-operative game interpretation can be depicted as follows. I chooses the row, II the column and III chooses between the two alternative matrices below:

Matrix 2

		II	
		a_2	b_2
I	a_1	0,0,0	1,0,1
	b_1	0,1,0	1,1,1

As long as the preferences are defined by each country independently of the others, on the basis of only two values, the introduction of additional players does not modify the structure of the prototype (see, for example, matrices 2 and 3).

The treatment of players' preferences will be the cornerstone of the following analysis and the key used to transform the prototype matrix in various games in order to reveal different types of decision-making crises. Progressive refinements in preferences will be successively introduced. In a first step, preferences are assumed to be independent, as in the games most often studied in game theory. Secondly, we shall turn to a form of interdependence among preferences which leads to four types of preferences. Finally, and as a third step, additional cognitive features relative to knowledge of preferences will be investigated in order to identify the kind of game the players are actually playing.

Matrix 3

		II	
		a_2	b_2
I	a_1	0,0,0	1,0,0
	b_1	0,1,1	1,1,1

2 FIRST LEVEL ANALYSIS: MYOPIC STRATEGIES AND INDEPENDENT PREFERENCES

At first glance, matrix 1 contains all the necessary ingredients to depict the normal form of a non-co-operative 2×2 game. Indeed, country I and country II are aware of all the rules of the game and their own preferences, and of the consequences of each of their possible actions, as far as the other's preferences are concerned. We should note that assumptions on knowledge of preferences are only derived from values 1 and 0. This point will be investigated in section 4 of the chapter. In all cases, according to Harsanyi's (1967–8) definition, the game thus understood belongs to the category of complete information.

This game has two interesting properties. First, each of the four possible outcomes is a Nash equilibrium point. Secondly, none of the possible actions provides a dominant strategy. Therefore, if we frame an international system summarized by Matrix 1 in a non-co-operative game according to the usual sense of the term, we already obtain some precisions about the meaning of crisis in the sense in which the concept is used in the following analysis. Roughly speaking, 'crisis' labels here a situation where the rules of the games block the whole process of rational decision-making. We should observe that there is no way for decision-makers to operate a discriminating choice and that all pairs of strategies are the best strategic replies, each of them *vis-à-vis* the other. This very special situation can be explained by the following characteristic. The outcome of each strategy depends only on the strategy chosen by the other player. For instance, if II chooses a_2, the payoff would be the same for I, whatever strategy he plays, when II chooses b_2 (and symmetrically for II's payoff if I chooses a_1 or b_1).

The anatomy of this crisis may be improved by analyzing the way to avoid such a situation, through transforming the initial game. We shall explore two different directions:

(a) on the one hand, according to Nash, threats can be introduced in the game. We propose to call it the 'deterrence prospect';
(b) on the other, according to Schelling's (1960) intuition, the knowledge of additional information can be assumed to be outside the game. We propose to call it the 'co-ordinate prospect'.

Extensive literature exists on deterrence models applied to international relations via game theory (Brams, 1985; Rudnianski, 1986, 1991; Powell, 1990). We shall therefore select some of the features of the deterrence approach relevant to our discussion of the relationship between payoff dominance and the crisis in the decision-making process.

Assuming a credible threat exists, a simple deterrence model can be introduced by modifying the initial interpretation of the two nuclear powers of the content of b_1 and b_2; this now means, respectively, according to the nuclear power interpretation previously suggested: 'Don't attack but respond'. If this is so, the original matrix 1 is transformed as in matrix 4.

Matrix 4

		II	
		a_2	b_2
I	a_1	0,0	0,0
	b_1	0,0	1,1

The threat which is then included in b_1 and b_2 drives the players to make the following commitment: 'If you attack, I shall retaliate'. Granted that full credibility is given to this commitment, the rational reaction of the other player eliminates the risk. Such a commitment corresponds precisely to the manoeuvre of deterrence in the field of nuclear strategy. In this game, b_1 dominates a_1, and b_2,a_2, in spite of the fact that the game has two equilibria, namely (a_1,a_2) and (b_1,b_2).[3]

In the above interpretation, a mutual and simultaneous commitment of this kind is implicitly assumed. This condition of similarity, which is obviously sufficient to guarantee that the equilibrium point (b_1,b_2) will be reached by the players in addition to the assumption of rationality, is not in itself mandatory. If, for instance, I plays first, he will then play b_1 because he knows that II will play b_2; which is what II actually will do, since I's self-commitment is credible and it is also assumed that he is rational.[4]

To sum up, a deterrence system is able to avoid a crisis in international decision-making as depicted in matrix 1 under the three following conditions:

(a) the deterrer and the deterred both act by referring to the dominance principle (rationality condition);
(b) the deterrer knows that his adversary acts rationally (knowledge condition);
(c) the deterrer's commitment is credible for the deterred (credibility condition).

We shall now investigate briefly the second direction for avoiding the crisis. If we return to matrix 1, we observe that the payoff vectors associated to the four equilibria points are not the same. Only one equilibrium, (b_1,b_2), is associated with a maximum payoff for both players (optimality). The actual difficulty for the players to select a strategy in order to reach this optimal equilibrium point then stems from a problem of co-ordination. Indeed, only (b_1,b_2) is a co-ordination equilibrium, as defined by Schelling (1960). Therefore, the solution of the crisis derived from matrix 1 can be reached through solving the co-ordination problem set by the 2×2 game.

An answer could be found with the help of the criterion of 'payoff dominance' promoted by Harsanyi and Selten (1988) and used here in order to co-ordinate the respective expectations of country I and country II *vis-à-vis* each other. Not only with pair (b_1,b_2) do the interests of I and II converge positively but this equilibrium point also satisfies Harsanyi's and Selten's dominance payoff criterion (Harsanyi and Selten, 1980, pp. 80–90). Therefore, according to the definition of Harsanyi and Selten, (b_1,b_2) is the most attractive equilibrium for both countries at the same time. The application of this rule appears to be both obvious and self-sufficient in solving the co-ordination problem which lies at the root of the crisis of the decision-making system, without changing the nature of the game. The question of co-operation provides a concrete interpretation of such a treatment of crisis in international fields, where the aim is to deduce a co-operative solution from a gaming structure which is *prima facie* non-co-operative, because of the lack of confidence among the countries concerned and the absence of enforceable agreements.

Unfortunately, things are not so easy.[5] The information set associated with the initial game does not allow to know whether country I and country II prefer to be satisfied independently, together, or have no preference between these two situations.[6] Therefore, the interpretation of payoff dominance previously suggested for solving the crisis requires an extension of the initial game, as well as for the previous deterrent solution.

On what grounds do country I and country II prefer equilibrium (b_1,b_2)? Assuming the initial definition of the preferences, the only answer seems to be found in placing, in the background of the game, the heroic hypothesis of the maximization of the payoff dominance criterion as 'common knowledge' (Harsanyi and Selten, 1988, p. 359). The intuitive signification of this hypothesis can be stated as follows:

(a) country I and country II accept the payoff dominance criterion;
(b) country I and country II know that they both accept the payoff dominance criterion;
(c) each country knows that the other knows that it knows, etc.... dominance criterion.

One way is to consider such a hypothesis as an essential ingredient of rationality, following Aumann (1976, 1987) and Harsanyi and Selten (1988). The criterion would therefore be inside the game. The realism of such an assumption appears highly questionable in the field of international relations between countries which are often very different.

An alternative meaning of this hypothesis of common knowledge, which is more in the spirit of Schelling (1960, pp. 294–8), would be to consider the payoff criterion not so much as a cognitive feature of the rationality implicitly assumed by the game but as something from outside the game. We should suppose, for instance, a common principle of 'fairness'. In this case, whatever the preferences of the two countries may be, (a_1,b_2) and (b_1,a_2) would be eliminated because of the unfair payoff associated with their outcome. Therefore (b_1,b_2) would be selected and the payoff dominance criterion would need no strengthening of the condition of rationality to be implemented.[7] Unfortunately, such a type of common knowledge between countries risks being as unrealistic as the previous one.

A conclusion can be drawn from this first level of inquiry: considering the prototype matrix as a traditional non-co-operative game does not provide any direct solution for selecting a best pure strategy on the rational basis of the payoff dominance,[8] as long as the assumptions of the game are not modified; therefore it would be more convenient to move on to a second level of investigation as regards deterrence, as well as co-ordination, characterized by some refinement in the definition of preferences.

3 SECOND-LEVEL ANALYSIS: INTERDEPENDENT PREFERENCES AND TYPES OF GAMES

The treatment of preference is the actual cornerstone in the chosen prototype of crisis elaborated in the non-co-operative game previously developed. We should recall that this is very roughly defined on the sole basis of the values which are allocated independently by each player to the different outcomes.

According to this convention, for country I as well as for country II, (b_1,b_2) is preferred to (a_1a_2), although the question we may ask is what country I's preference would be between (b_1,b_2) and (a_1,b_2) on the one hand, and between (a_1,a_2) and (b_1,b_2) on the other. This same question may be raised concerning country II's preference between (a_1,a_2) and (a_1,b_2) as well as between (b_1,b_2) and (b_1,a_2). Two answers are possible: either these outcomes are considered as equal by the players, or there is a lack of comparison between them due to the absence of sufficient information for the players.

These two interpretations have neither the same meaning nor the same consequences. According to the former, the two countries are, in fact, indifferent to the other's evaluation of the outcome. For instance, the preference order of country I among the possible outcomes is to be written as follows: (a_1,b_2) E (b_1,b_2) P (a_1,a_2) E (b_1,a_2) where E symbolizes a relation of equivalence and P a relation of strong preference. Thus, an order of preference does exist for each player, but it does not allow the players to discriminate a best strategy. According to the latter, there is no way for the countries to order the set of all the consequences of their actions. In such a case, it can be argued that the prototype depicted by matrix 1 cannot work as a game according to its traditional meaning in game theory, because of the absence of an essential feature of any one game, namely the definition of a complete order of preferences on the set of all the outcomes associated with each player. No one is actually satisfactory.

Therefore we now propose a definition of the interdependence of the preferences which involves a change in their field of interpretation. Each player ranks directly the four payoffs according to his own preferences; but the preferences of each player are based on the knowledge he possesses of the other player's evaluation, provided by the payoff associated with all outcomes. We should note that such types of preference do imply the necessity for each player to be aware of the other's order of preference.[9]

Four types of these preferences can now be considered. All are 'rational' in the sense that they are consistent with the conventional meaning attached to 1 and 0. This kind of minimal rationality only excludes 'masochistic behaviour'. They correspond to four different strong orders in the set of all possible outcomes listed below:[10]

P1 = (1,0) > (1,1) > (0,0) > (0,1)
P2 = (1,1) > (1,0) > (0,1) > (0,0)
P3 = (1,0) > (1,1) > (0,1) > (0,0)
P4 = (1,1) > (1,0) > (0,0) > (0,1).

P1 and P2 correspond to two 'polar' cases, respectively inspired by 'malevolence' and 'benevolence' towards the other player. The first case most often occurs in the field of military affairs, whereas the second can be linked to utilitarian ethics more often relevant in environmental affairs. P3 and P4 reveal two different kinds of mixed attitudes which can almost always be found in matters of diplomacy and also in international economic relations.

The information provided by matrix 1 can now offer a general pattern for sixteen different 2 × 2 non-co-operative games, following the total combination of these four orders of preference. Such a property explains why matrix 1 has been given the character of a prototype of crisis in decision-making systems. Indeed, a typology of crisis systems emerges from the study of the different games, all being consistent with matrix 1.

In order to present the various configurations, let us start gathering all 2 × 2 games in which the players have the same kind of preferences. 4, 3, 2, 1 are ordinal indices associated respectively with first, second, third and fourth rank in the preference order. The normal form of four corresponding games with symmetric preference orders for both players are illustrated by the matrices for G_1 to G_4.

Both G_1 and G_2 have only one equilibrium point, which corresponds to a pair of dominant pure strategies. The difference is that the equilibrium point (a_1,a_2) in G_1 is not Pareto-optimal, while the equilibrium point (b_1,b_2) in G_2 is Pareto-optimal. Both G_3 and G_4 have two equilibria points, neither of which correspond to a pair of dominant pure strategies. In G_3, the player's interests are almost divergent in the two equilibria points, so each player prefers one of them (and not the same). Conversely, they are convergent in G_4's two equilibria points. Three of the four games are well-known figures in game theory, named in the literature as 'Prisoner's Dilemma' (G_1), 'Chicken' (G_3) and 'Stag Hunt' (G_4). Then, the same objective conditions, according to the players' own interests, which were summarized by the prototype matrix 1, generate different types of anomalies in relation to the systems of preference of the two players. An analysis of such anomalies offers a starting point for the classification of crisis in the underlying decision-making process.

II

I		a_2	b_2
	a_1	2,2	4,1
	b_1	1,4	3,3

G_1

P1,P1

II

I		a_2	b_2
	a_1	1,1	3,2
	b_1	2,3	4,4

G_2

P2,P2

II

I		a_2	b_2
	a_1	1,1	4,2
	b_1	2,4	3,3

G_3

P3,P3

II

I		a_2	b_2
	a_1	2,2	3,1
	b_1	1,3	4,4

G_4

P4,P4

If the notion of crisis completely disappears in the case of G_2 which is relatively rare in the dominant trend of *Realpolitik*, can we still speak of a crisis characterizing the decision-making process in G_1, in G_3 or in G_4? From the formal point of view in game theory, there can be no doubt that the Prisoner's Dilemma is not a surprise inasmuch as there is one pair of dominant strategies (a_1,a_2) and that each corresponds to the best strategic response and is, thus, consistent with the Nash definition of equilibrium. A doubt can however persist in the players' mind, generated by backward induction, when the outcome of the equilibrium point coincides with common disaster (see matrix 1). We should observe that backward induction applied to a repeated game of the Prisoner's Dilemma leads to the same result however many times the game is repeated (Selten, 1978). In this case, the crisis merely designates a perverse effect in the decision-making process. Therefore, non-myopic strategies are to be explored in order to avoid the common disaster. Thus the deterrence model can be re-examined in the light of a sequential game of two players with hostile preferences (Rudnianski, 1991).

Another example, slightly different from the prototype, demonstrates that such a perverse effect can be observed in a case different from that of the Prisoner's Dilemma. We refer to the so-called Gulf Crisis which preceded the Gulf War at the end of the summer of 1990. Looking at the relationship between Iraq and the United States after the Iraqi invasion of Kuwait, schematically each country had two options at its disposal: 'withdraw' or 'stay' in Kuwait, as far as Iraq was concerned, and 'increase' or 'decrease' the military pressure, for the United States (see matrix 6).[11]

Matrix 6

		USA	
		increase	decrease
Iraq	stay	0,0	1,0
	withdraw	0,1	0,1

Framing in the scope of a classical non-co-operative game corresponding to the first level analysis, the game has three equilibrium points including 'stay in Kuwait' for Iraq and 'increase the military pressure' for the United States, though the latter do not have any dominant strategy. We can reasonably assume for Iraq, as well as for the United States, a profile of hostile preferences (P1). Hence the game can be shaped in the normal form; where I is Iraq and II the United States.

		II	
		a_2	b_2
I	a_1	2,2	4,1
	b_1	1,4	1,4

G_5

Therefore when, in addition to the logical values (matrix 6), a relevant system of preferences (G_5) is taken into account, the only non-co-operative equilibrium is now (a_1,a_2), which means staying in Kuwait for Iraq, on the one hand, and increasing military pressure for the United States, on the other. Thanks to the proposed general framework, it becomes clear (a) that the existence of many equilibria due to the configuration of interests in a given 2×2 situation does not necessarily lead to a crisis of indetermination in the decision-making process; and (b) that, conversely, the existence of a unique equilibrium point can lead to the consequence that both parties are penalized. Note that the same result can sometimes be reached with other preference orders. This is the case in the Iraqi example: as soon as the United States knows that Iraq can be considered rational in the very broad sense associated to the meaning of the logical values (Schmidt, 1991a).[12]

Let us now move to G_3 and G_4 in which the associated crisis has quite a different meaning than in G_1. In both cases, two characteristics are combined: the existence of two equilibria, on the one hand, and the impossibility for a player to choose among his available actions according to his preferences on the basis of the criterion of dominance. We can therefore argue that none of these equilibria points is accessible for a player (Schmidt, 1991b) or, alternatively, that there is still a problem of equilibrium selection (Harsanyi and Selten, 1988). An equilibrium point is labelled as 'accessible' if the information available to each player and the assumed rational conditions are sufficient to guarantee the choice of the corresponding strategies. The existence of several equilibria does not necessarily lead to a problem of selection when only one of the equilibria is accessible. This is the case, for example, in a two-player game with only one pair of dominant strategies (see Schelling, 1960, p. 297). On the contrary, the unicity of an equilibrium point does not guarantee its accessibility (Harsanyi, 1977, p. 125).

The difference between the Chicken crisis and the Stag Hunt in the decision making process is underlined by the tentative application of the two criteria proposed by Harsanyi and Selten, in order to select one of the multi-equilibria in a non-co-operative game. Their purpose can be summarized by a quest aimed at strengthening the dominant principle as a foundation for the selection of the best strategic answer for both players. Then they propose two sub-criteria, labelled respectively 'payoff dominance' and 'risk dominance' (Harsanyi and Selten, 1988, pp. 75–85).

Let us first of all apply them to G_4. The former, which has still to be defined, leads to the selection of the equilibrium point (b_1,b_2). The latter, which intuitively coincides with the pair of risk-free strategies, leads to the selection of the other equilibrium point (a_1,a_2). Thus, the case of G_4 illustrates the logical possibility of a contradiction between the two sub-criteria of dominance elaborated by Harsanyi and Selten, so as to formulate a general solution to the crisis problem induced by the combination of many Nash equilibria and no dominant pure strategies for the players. As already demonstrated, this contradiction can be removed by the strong additional assumption that payoff dominance is common knowledge. Applied to G_3 neither of the two criteria operates because the game has no payoff dominance and the risk-dominance does not lead to an equilibrium point. In the meanwhile the crisis in G_4 appears to be investigated through a refinement of individual rationality, the crisis in G_3 requiring other intellectual tools.

An exhaustive study would require the checking of all combinations of different preference profiles previously listed. Thanks to the symmetry between players in rows and players in columns, all the cases where players have different preferences can be summarized by games G_6, G_7, G_8, G_9, G_{10} and G_{11}.

Three observations emerge after a quick examination of these six games. First, none of the six games has more than one equilibrium, which means that the form of crisis linked to a problem of discrimination among equilibria disappears when the players in the game have different types of preferences. This observation answers the previous remark concerning the games where the players have, on the contrary, the same types of preferences (G_1, G_2, G_3, G_4) and where such a form of crisis appears twice in G_3 and G_4. Secondly, when one equilibrium point does exist, it always corresponds to a dominant strategy for at least one player. However, one cannot conclude too rapidly that the equilibrium is accessible for both players and that various difficulties on this ground which were pointed out in the games previously studied also disappear. Thirdly in one of the games, namely G_{11}, there is no equilibrium point, at least when defined as a pair of pure strategies. Such a possibility opens up room for a new kind of crisis which will be commented upon briefly.

Let us now focus our attention upon the cases where one of the players, say country I, has a hostile preference order *vis-à-vis* the other, say country II. The reason why these cases have been chosen for further elaboration is because they occur in international conflicts where, most often, one country at least has such a preference order.

Indeed, there is no crisis at all in the decision-making process framed by G_6. This game only has one equilibrium point (a_1,b_2) corresponding to a pair of dominant pure strategies. For G_7 and G_8, a_1 is the dominant pure strategy of I, and II has no dominant strategy. However, II, knowing that I moves rationally, will choose b_2 in G_7 and a_2 in G_8. Such observations, added to the fact that in two of the games, namely G_6 and G_7, the payoffs of the outcome corresponding to the equilibrium point (a_1,b_2) are most satisfactory for the aggressor, would appear to justify the so-called 'offensive priority', which goes back to Clausewitz. They also support a kind of rational justification in favour of a hostile preference profile, frequently observed in the life of international relations between the two countries.

Finally, the combination of the two types of intermediate preferences between 'malevolence' and 'benevolence', namely P3 and P4, offers, as we have seen, the opportunity to introduce another and final kind of crisis into the decision-making system. G_{11} has no equilibrium point corresponding to a pair of pure strategies. An equilibrium point could obviously be found by means of the definition of a mixed strategy. However, the interpretation which must be provided to support such mixed strategies is controversial in the reality of international life.

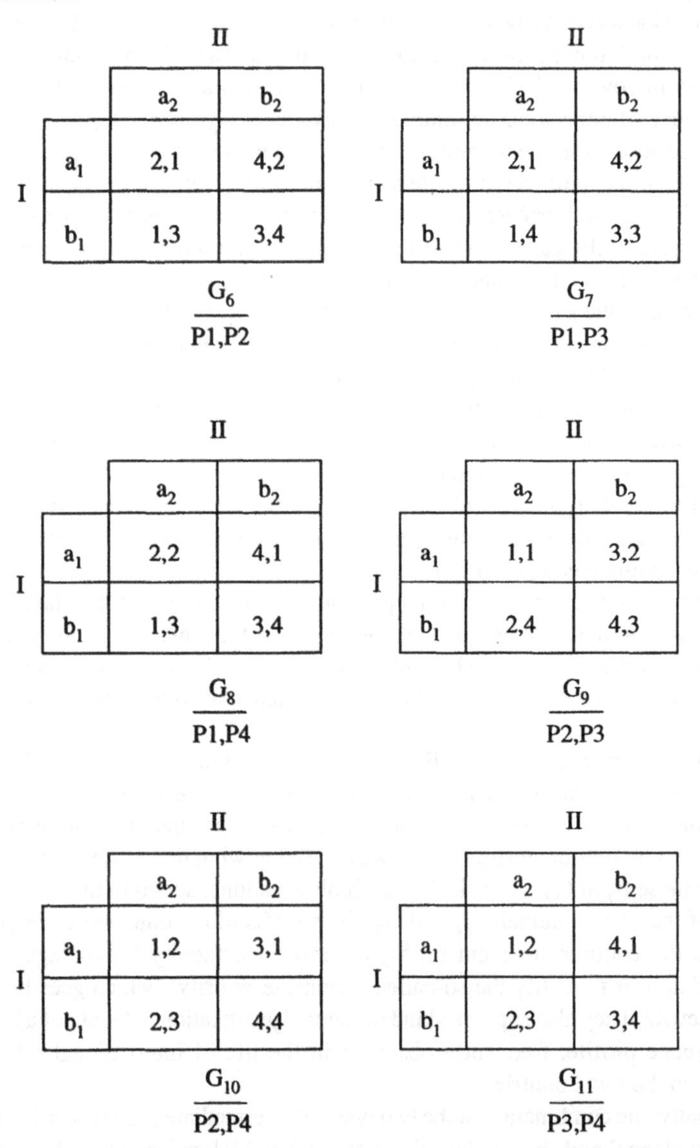

I \ II	a_2	b_2
a_1	2,1	4,2
b_1	1,3	3,4

G_6

P1,P2

I \ II	a_2	b_2
a_1	2,1	4,2
b_1	1,4	3,3

G_7

P1,P3

I \ II	a_2	b_2
a_1	2,2	4,1
b_1	1,3	3,4

G_8

P1,P4

I \ II	a_2	b_2
a_1	1,1	3,2
b_1	2,4	4,3

G_9

P2,P3

I \ II	a_2	b_2
a_1	1,2	3,1
b_1	2,3	4,4

G_{10}

P2,P4

I \ II	a_2	b_2
a_1	1,2	4,1
b_1	2,3	3,4

G_{11}

P3,P4

Finally, a comparison between G_{11} and M_1 is most interesting. In both cases, there is no dominant strategy. Nevertheless all the outcomes are equilibria in M_1 and there is none in G_{11}. Their comparison suggests first of all that instability and uncertainty are not necessarily linked. Indeed, international systems corresponding to G_1 and G_{11} are both uncertain. However, uncertainty is closely related to instability in the case of G_{11}, as has been demonstrated by the cyclical process of a sequential game. This is quite the contrary in M_1 where the corresponding system is 'self-stabilizing', in the sense that whatever outcome is obtained, no player has an incentive to move. Furthermore, full stability as in M_1 as well as full instability as in G_{11} lead to two opposite kinds of crisis.

To sum up, according to this second-level analysis the prototype of the combinations of interests depicted by the prototype matrix is consistent with various types of games. Only some of them can generate a crisis in decision-making which is directly derived from the players' system of preferences. The games in which the players have the same preference order most often lead to a crisis in decision-making, except for the special case where the players have different mixed type of preference (G_{11}), which does not allow any pure strategy equilibrium. Some crises are due to a lack of dominance in the player's set of pure strategies (G_3, G_4), others to the undesirable consequences of dominant strategies (G_1).

Assuming that the system of preferences of a country is a basic component of an international decision-making crisis, a new problem emerges concerning the knowledge of each country *vis-à-vis* the preferences of the others. This question is almost always included in the assumption over rationality, because of the traditional definition of independent preferences in game theory. This reduction is not yet possible since our definition of preferences sets forth four types of rational preferences. Ignorance about the true system of preferences implies here an uncertainty regarding the true game which the players are playing and this can also generate a crisis in decision-making. We shall now examine some of the implications of knowledge of the other's preference on the rational choice in a decision-making system stylized by the prototype matrix.

4 THIRD-LEVEL ANALYSIS: KNOWLEDGE ON PREFERENCES AND IDENTIFICATION OF THE TRUE GAME

Starting from a general canvas provided by the prototype model given by matrix 1, we have introduced definitions of preferences in order to identify the type of game in question. Thus international crises have been apprehended by referring to two dimensions: the configuration of a country's interest, on the one hand, and its preference ordering, on the other. We implicitly assumed in section 3 of this article that each country knows the other's system of preferences. In other words, the international crises previously pointed out took place in a world of complete information, according to Harsanyi's (1967–8) definition. We shall now focus our examination on the information known by each player about the other' s system of preferences in order to introduce games of incomplete information.

We should first observe that complete information does not necessarily entail that the country's preferences are common knowledge. Thus country I, for instance, can know country II's preference, though it ignores that country I knows that it knows its preference (and the same symmetrically for country II *vis-à-vis* country I's preferences). Common knowledge is then intuitively understood as a strengthening of the hypothesis of complete information. The introduction of preferences as common knowledge may have different consequences on the existence of the crisis previously identified according to the game under consideration. It does not change the understanding of the crisis labelled 'perverse effect', as demonstrated by G_1 and G_8. Conversely, some of the 'undecidability crises' can disappear when preferences are common knowledge, as in G_4, because they imply a payoff dominance as common knowledge (see above) whilst this is not the case for others, as in G_3.

In any case, the empirical relevance of complete information on preferences looks very unrealistic in the field of international relations for three major reasons, at least. First, the countries' preference order is collective and most often derived from a compromise of opposing viewpoints inside the country and therefore difficult to predict; secondly, governments do not like to reveal their true preferences for obvious reasons; thirdly, the countries' preferences cannot be easily induced from past political choices, such as those in the economic sphere, because of the general lack of constant co-ordinates, such as prices and quantities, in the microeconomic analysis of consumption. Therefore we have to abandon this assumption on the grounds of realism. But the immediate consequence is an additional difficulty for the countries in knowing what game they are playing and, then, to what kind of crisis they are exposed.

In order to illustrate the kind of situation which can lead to the relaxation of the assumption about complete information on preferences, let us suppose that country I has a P1 preference order and country II a P3 preference order. The preference system is then P1,P3 corresponding to G_7. We must now assume that country I does not know country II's preferences, but that, conversely, country II knows country I's preferences (asymmetrical information structure). This means that country II knows the true game, that is G_7, which is ignored by country I.

Viewed from country II, its knowledge of the preferences of country I is not sufficient information to guarantee that the game it actually has in mind is the *true* game. Additional information about country I's knowledge on its preferences is necessary. Three cases then appear. Country II knows that country I ignores them, or it believes wrongly that country I knows them, or it does not know anything about the question. In the first case, country II has to imagine all the games consistent with country I' s bounded information, namely G_7 but also G_3, G_9 and G_{11},[13] before choosing his move. In the second case, it will wrongly play the true game and be exposed to the risk of an unexpected perverse consequence of a decision computed, however, as rational. Suppose, for example, that country I chooses its strategy on the grounds of G_9; it will then choose b_1 and the outcome will be b_1,b_2 which is not an equilibrium in the true game. In the third case, it is alleged to have reshaped the situation in a Bayesian framework (Harsanyi, 1967–8, 1977), then used an *a priori* probability of distribution regarding country I's best strategy in the four possible games. In none of these three cases can country I and country II be sure of playing correctly the true game.

Thus now neither country I knows country II's preferences, nor country II knows country I's preferences (case of symmetrical ignorance). Neither of them, consequently, knows the true game, and the set of computable information for country I is not the same as that for country II. Assuming the very weak rationality as common knowledge, only four games are to be taken into account by country I with Pl preference, namely G_1, G_6, G_7 and G_8. Each of the four games have only one equilibrium point corresponding to a dominant pure strategy for country I in every game. Furthermore, and fortunately, this dominant strategy a_1 is the same in all the possible games. Country I will then choose a_1 without any preconceptions.

As far as country II with P2 preference is concerned, the four games to be taken into consideration are G_3, G_7, G_9 and G'_{11} which have already been checked. However, the choice of an action is much more complex for country II than for country I. First of all, neither of the two available pure strategies is dominant in any of the four games. Furthermore, the existence and number of equilibrium points associated with a pair of pure strategies vary from one considered game to another. If G_7 and G_9 have only one equilibrium (which is not the same), we should recall that G_3 has two equilibria and G'_{11} has none at all. Last but not least, b_2 leads to Nash equilibria in G_7, a_2 in G_9, and a_2 to one equilibrium and b_2 to the other in G_3. Yet the whole of country II's available information does not allow it to select a pure strategy on the grounds of Nash's definition of equilibrium and dominant pure strategies.

A static interpretation does not seem sufficient to solve the difficulty, but a dynamic one can be imagined in the following terms.[14] According to previous observations, strategy a_1 dominates strategy a_2 in every relevant game for country I, while country II does not have sufficient information to choose its strategy rationally; one can therefore expect that country I moves first in a_1 because his information is more efficient than country II's information. Country I's move provides country II with the assumption that it acts rationally, which is a sufficient basis to eliminate the possibility of G_9 and G'_{11}. Thus, country II will choose a strategy which will be consistent with G_3 and G_7, namely b_2. Being aware of country II's move, country I in turn will now eliminate G_1 and G_8 on the same grounds as for its preceding move. Indeed, the dynamic *tâtonnement* just sketched out converges to (a_1,b_2), which is one of the two Nash equilibria in G_3, and the only one in G_6 as well as in G_7 (the true game). Following such a process of *tâtonnement*, the initial problem is actually solved and, in the end there is no crisis of the kind previously discussed in the decision-making system, even if at the same time neither of the two countries knows precisely in which game it is taking part.

The general procedure which has been followed here can be then summarized as follows. First, the initial set of information on the preferences of the players provides material to frame the situation as a hypergame. In the case studied here, G_1, G_3, G_6, G_7, G_8, G_9 and G'_{11} are the components of the hypergame denoted Γ. The relevant domain of interpretation is composed of G_1, G_6, G_7 and G_8 for country I's definition of strategy and by G_3, G_7, G_9 and G'_{11} for country II's definition of strategy. Secondly, the players' moves are utilized as a procedure to discriminate between the components of the hypergame in order to obtain a more tractable one. At the end of the procedure in the chosen example, the final hypergame Γ' comprises only G_6, G_7, G_3. The essence of the method can be stylized as in Figure 5.1.

Figure 5.1

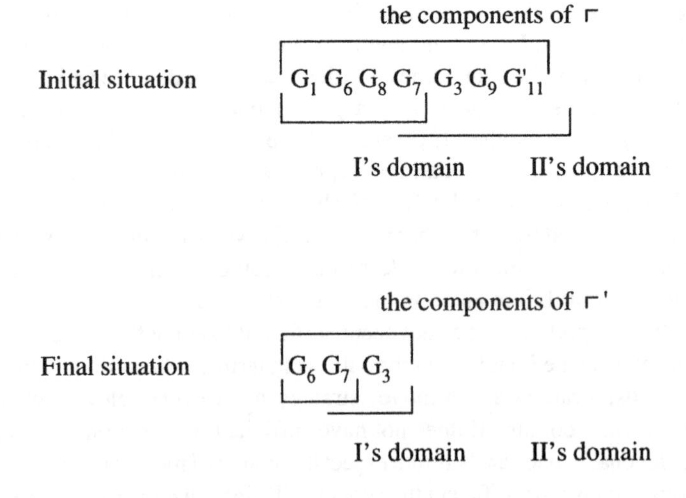

It may be argued that this example refers to an *ad hoc* situation, from which no general properties can actually be deduced. Indeed, the proposed procedure does work in the example because (a) it converges to only one equilibrium and, (b) the pair of strategies (a_1,b_2), corresponding to the equilibrium point, is accessible to the players via a kind of process of *tâtonnement*, mixing together beliefs on rationality and forward induction.[15]

We must now briefly survey all the other cases where at least one of the two players has a P_1 type of preferences and where neither of the two players knows the other's preferences.

When country I and country II have both P_1 preferences, b_1 dominates b_2 as well as a_1 dominates a_2 in all the relevant games. Then the equilibrium point (a_1,b_1) which is reached by the players is the equilibrium of the true game G_1, namely the Prisoner's Dilemma. The process of *tâtonnement* does not work in this case because the knowledge of the other's move does not provide any additional information and, for this reason, is of no interest to the players. Incidentally, this explains the very extended domain of the Prisoner's Dilemma type of crisis.

The situation is almost the same when country I has P_1 preferences and country II P_2 preferences. In this case, b_2 dominates b_1 in every relevant game. As a_1 dominates a_2, the equilibrium reached is (a_1,b_2) which is also the equilibrium of the true game (G_2) and the *tâtonnement* has no *raison d'être* in this case either.

When country I has P_1 preferences and country II P_4 preferences, we return to the same kind of situation as the example at the beginning. Player I can move first because his strategy is dominant in every relevant game. This move provides country II with sufficient information to eliminate two possible games and leads it to choose rationally a_2. Thus, as in the case with P_1 and P_3, the players reach the equilibrium point of the true game G_8 (a_1,a_2) without knowing precisely in which game they are taking part.

We have chosen to investigate all possible cases where country I has P_1 preferences, because we assume that, in the field of international crisis, the attitude of one country, at least, is usually hostile *vis-à-vis* the other (see below G_1, G_6, G_7, G_8). Three main points emerged from the study of this reduced sample. First, the ignorance of each player regarding the other's preferences does not entail any crisis as long as the true types of preferences of the other are 'malevolent' or 'benevolent' according to our previous interpretation in section 3 (P_1,P_1 and P_1,P_2). On the contrary, difficulties arrive when its order of preferences belongs to a mixed type (P_1,P_3 and P_1,P_4). In this case, a process of *tâtonnement* always converges at the equilibrium point of the true game, thanks to the existence of a dominant strategy in all relevant games for the player with P_1 preferences. This observation seems to reinforce the previous notation on the kind of rationalization of a hostile preference profile among the nations (see below).

Finally, we must stress that the convergence of the process of *tâtonnement* which has been observed in our sample cannot be taken as a guarantee. Let us consider the case where country I has P_3 and country II P_4 preferences. Not only the true game, G'_{11}, has no pure strategies equilibrium but the *tâtonnement* described above does not work. The crisis in the decision-making process is then still aggravated.

Another kind of crisis in decision making can be seen from this investigation of a player' s knowledge of the other player's preferences. According to the hypergame approach, we have noticed that the sets of information associated with each player have only one intersection, corresponding to the true game. It means that the world of country I is largely different from the world of country II and opens the door to another type of crisis, where they become completely disconnected. Such an opportunity can arrive when the countries in question believe wrongly that they know each others' preferences (see above, case 2).

In order to give a picture of such a situation, let us examine the case where country I with a P_1 preference order believes that country II has a symmetric preference order, whilst country II, with a P_3 order, believes that country I has the same. Hence country I believes it is playing in G_1, that is a game of the Prisoner's Dilemma, and country II that it is playing in G_3, a Chicken game. G_1 and G_3 must then be understood as subjective 'mental games' in the sense that both are the result of subjective framing of the same prototype situation on the grounds of spurious assumptions. None of the players takes into account the true game, G_7.

According to the representation of the Prisoner's Dilemma, country I will choose a_1, which is its dominant pure strategy and its best strategic response to country II's strategy, which it had forecast erroneously. The decision is far more difficult for country II, but one can reasonably expect that it will act quickly, choosing a_2 which gives it a chance to reach the more favourable of the two equilibrium points in the Chicken game to which it is referring. The pair of strategies rationally chosen by country I and country II (a_1,a_2) on the basis of their subjective games is not an equilibrium point in the true game G_7, where the equilibrium point of unique pure strategies is (a_1,b_2). Furthermore, it leads to a very bad situation for both of them.

A more detailed analysis shows that the first move here determines the final issue, when each country is not quite sure of its assumption regarding the other's own preferences and could not even ascribe a probability value to its belief. If country II moves first and chooses a_2, according to the rational prescriptions of the Chicken game, it even reinforces the determination of country I to choose as its strategy a_1, because a_1 dominates a_2 in G_1, as well as in the 'true' game actually played. If country I moves first, it then also rationally chooses a_1; the best strategic response for country II is no longer a_2 but b_2, because b_2

now dominates a_2 in G_3 as well as in G_7. The equilibrium point of the true game will then be reached in spite (or perhaps because) of the countries' complete ignorance of the game they are actually playing[16]. But why does country I move first, when country II has a stronger incentive to rapidly choose a_2? Information from outside sources could then provide additional features inflexing the decision-making process (cf. Schelling's focal points).

An illustration of this kind of crisis, which we propose to label 'misperception crisis', can be found during the well-known Cuban Missile Crisis in 1962. An interpretation of the main options available for the two protagonists in the prototype model can be proposed where a_1 and b_1 respectively mean 'keep the missile bases in Cuba' and 'withdraw the missile bases from Cuba', for the Soviet Union, and a_2 and b_2 air-force attack Cuba' and 'boycott Cuba', for the United States (see matrix 7).[17]

Matrix 7

US

		attack a_2	boycott b_2
Soviet Union	Keep a_1	0,0	1,0
	Withdraw b_1	0,1	1,1

Thanks to the prototype framework, without any additional assumption on preferences, the decision-making system clearly appears blocked (first level of an 'undecidability crisis'). Therefore, there was normally a pressing temptation for the two countries to elaborate quickly these additional assumptions on preferences, in order to provide each side with a framing reference for decision-making (second-level analysis). Without completely adhering to the dominant thesis of the Chicken game interpretation, it appears plausible that, at least, some of the US President's advisers considered the situation as a G_3 Chicken game (Allison, 1971).[18] On the other hand, we can imagine that Soviet decision-makers were probably more inclined to shape the situation as a G_1 Prisoner's Dilemma, according to our retrospective belief about their own military preferences. Therefore, on the one hand, there was pressure on the American side to decide on a rapid air-force attack by reference to the G_3 subjective prospect whilst, on the other, a logical calculation led the Soviet government to maintain their missile installation in Cuba. A conjunction of these two rational decisions inside two different appraisals of the situation, both consistent with available data and the interests of these powers, would then have risked transforming the initial 'decisional crisis' into a catastrophic war (a_1,a_2), through a 'misperception crisis' as described above.

Why did the two countries finally escape from this pessimistic but paradoxically non-irrational negative scenario? We should remember that, in framing the prototype model which depicted the features of the situation, in a definite game each country assumed that it knew the other's preferences, and had simply projected its own preferences on the other. The beginning of a suspicion about the validity of this belief, following the direct contact between Kennedy and Khrushchev during the thirteen days of the 'decisional crisis' offers a tentative explanation. It did not mean rejecting the prototype model but was aware of the cognitive difficulty of framing the situation in a definitive game, supported by a firm assumption of the other's preferences. Therefore, the two powers, progressively conscious of their lack of adequate information, returned to the initial formulation in matrix 1 and tried to find the solution in another direction which corresponded to the analysis of a problem of co-ordination (see Schmidt, 1991a, pp. 162–7).

In principle, such a difficulty might occur every time each player, whatever his preferences, wrongly evaluates the other player's type of preferences. Fortunately, the set of cases relevant for our study can be drastically reduced. Obviously, the main temptation for each decision-maker is to lend his own preference to others. Therefore our investigation will be bound to that category. Furthermore, we assume here that in the area of international relations, at least one of the countries has a P_1 preference, whilst a P_2 preference is too rare to be taken into account. Therefore outside the situation previously examined where the true game would be G_7 and (P_1, P_3) the pair of preference orders, we can restrict our attention to another case, that where the true game would be G_8 and (P_1,P_4) the pair of preference orders.

Country I's subjective game is, as before, also a Prisoner's Dilemma, while the subjective game of country II is now a Stag Hunt. If country I moves first and chooses a_1 for the reason previously mentioned, country II will choose a_2. Therefore, they will both reach (a_1,a_2), which is the only equilibrium in G_1, as well as one of the two equilibria of G_4. Furthermore, (a_1,a_2) is also the only Nash equilibrium in G_8, the true game being completely unknown to the two players. This example demonstrates that such a misperception on the other's preferences does not necessarily lead to a crisis in the decision-making system. It also illustrates again the cogency of the Prisoner's Dilemma in the field of international relations.

Finally, we would suggest a tentative interpretation of the recent history of US/Soviet strategic relations. The disappointment at the end of the 1970s following the SALT arms negotiations between the US and the Soviet Union could eventually be understood as an empirical illustration of this case. Indeed, the Americans imagined that they were playing G_4 during the period 1972–9, whilst the Soviets were always thinking of playing G_1. Therefore the arms race did not really stop.

CONCLUSION

The concept of a Nash equilibrium has been used as a guideline for the understanding of a class of international crises provoked by different kinds of spurious and often unexpected effects in rational decision-making processes present in international life. Two directions have been explored. The first tries to characterize such phenomena in connection with: (a) the absence of a pure strategic equilibrium (crisis of *instability*); (b) the existence of many equilibria without dominant pure strategies (crisis of *undecidability*); (c) a non-'satisfactory' payoff associated with the outcome of the equilibrium (crisis of *perverse rationality*). The second tries to identify the origin of the listed types of crisis in relation to different interpretations of a Nash equilibrium. Each of them is developed by reference to three levels of understanding which can be ranked by growing degrees of complexity, as follows: at the bottom, a general configuration of the countries' interests (the prototype matrix); then this configuration of interests plus the definition of the systems of preferences (the set of all possible games); and, at the top, the configuration of interests and the preference orders, plus the cognitive conditions of their knowledge (the framing process of the game played). One or another type of crisis can be observed at each of these three levels. Thus, thanks to Nash equilibria, crises can be detected in a definite game, as well as in a dynamic quest for the actual game or, even, of the general structure overlapping the possible games (the hypergame).

Some preliminary results emerge from this investigation. An 'undecidability crisis' is immediately derived from the first level, due to the property of equilibrium of all the outcomes of the prototype. The system of preferences determines the existence and the types of crisis, in a definite game (see for instance G_1, G_3, G_4, G_{11}). The ignorance of the game that countries are playing does not necessarily lead to a crisis, the advent of which is only dependent on the payoff values attached to the equilibria of the games belonging to the sub-set of the bounded information of the countries. On the other hand, some crises cannot be ascribed to a definite true game but to its misinterpretation by the players. However, and contrary to an intuitive point of view, such a crisis is not always insoluble, especially when one of the players has a hostile preference *vis-à-vis* the other.

Finally, the choice of the relevant level for the crisis analysis is obviously related to empirical data. Case studies seem to demonstrate that decision-makers can shift from one level to another during the real time schedule of a concrete crisis, as the re-examination of the Cuban crisis demonstrates. Therefore, improvements in the mechanism of the understanding of crisis must be expected from the empirical tests in the proposed general framework by means of detailed and carefully analysed case studies on its grounding.

NOTES

1. We assume in this paper that the international decision-making process concerns related choices by a unique category of decision-makers in each country, making these choices only on the grounds of international considerations. Such an over-simplification can be made more complex by the addition of supplementary players in each country (see Cederman's contribution, ch. 3 in this volume).
2. On condition that we consider only one strike for each country.
3. With a difference in their respective stabilities. The pair of pure strategies a_1,a_2 is not dominated by others, whilst b_1,b_2 dominates the others (cf. Kohlberg and Mertens, 1986).
4. For a study on conditions of credibility in a deterrence game, see Rudnianski (1986, pp. 133–4).
5. A crucial difference must be introduced between the player's pragmatical viewpoint and the model builder's semantical one. The background of this distinction up to Carnap is analysed in Schmidt (1991b).
6. Note that the concept of 'indifference' which is familiar to economists has two different meanings which are unfortunately confused on many occasions. It is not the same thing to judge two outcomes as equivalent as being unable to rank one *vis-à-vis* the other on a scale of preferences. Indifference designates a relationship of equivalence in the first case and a lack of a relationship of preference in the second.
7. There is a radical difference between Aumann's definition of common knowledge (1976, 1987) used by Harsanyi and Selten (1988) and the Schelling approach to the same idea. As we have already seen, Aumann considers common knowledge as a property attached to the assumption of rationality inside the game theory itself. On the contrary, for Schelling, the content of common knowledge is to be found outside the game operating as a 'hint' in a contextual system (Schelling, 1960, p. 295, n. 9). So when, for Aumann, common knowledge is no more than a cognitive complement of mathematical rationality, it must be captured out of the mathematical configuration of the payoffs for Schelling. Therefore, Schelling often speaks in this connection of 'another type of rationality'. These two semantic components of the concept of common knowledge often lead to opposite solutions.
8. The rational property questioned here is the 'principle of dominance', which is one of the theoretical pillars of rational choice (see Kahneman and Tversky, 1979, pp. 265ff.).
9. The formula 'external preferences', referring to the externality of this domain in the economic sense, would perhaps become a convenient label for such kinds of preferences, as opposed to the 'internal preferences' initially assumed in the prototype. For a slightly different treatment of interdependent preference by means of utility functions, see Nicholson, ch. 4 in this volume.
10. Other types of external preferences could also be studied by substitution of the strong orders, $\geq$, by weak orders, $>$ with all of the mixed combinations.
11. Assuming that military option was decided by the US government.
12. The extended form of the game is most illustrative. See Schmidt (1991a, p. 236).
13. On the condition of permuting the rows and columns for G_{11}.
14. In the sense of an 'evolutive' versus an 'eductive' interpretation of a Nash equilibrium discussed by K. Binmore. According to Binmore's definition, the evolutive interpretation is generated through a myopic *tâtonnement* (Binmore, 1990).
15. By 'forward induction', we mean a process of reasoning where players use further computation of information provided by the previous sequences of the game. With regard to the so-called opposition between forward and backward induction, see Kreps, 1990, and Schmidt, 1993b.

16. Assuming that both are aware of their own ignorance.
17. For a justification of the values associated to the outcomes, see Schmidt (1991a).
18. For a rejection of this hypothesis, see Brams (1977, and 1985, esp. pp. 53–4).

REFERENCES

Allison, G. (1971), *Essence of Decision: Explaining the Cuban Missile Crisis*, Boston, Mass.: Littlebrown.

Aumamn, R. (1976), 'Agreeing to Disagree', *Annals of Statistics*, **4**.

—— (1987), 'Correlated Equilibrium as an Expression of Bayesian Rationality', *Econometrica*, **55**.

Axelrod, R. (1984), The Evolution of Cooperation, New York: Basic Books.

Binmore, K. (1987–8), 'Modelling Rational Players I and II', *Economics and Philosophy*, **3** and **4**.

—— (1990), *Essays on the Foundations of Game Theory*, Oxford: Basil Blackwell.

—— and Dasgupta, P. (1986), *Economic Organization as Games*, Oxford: Basil Blackwell.

—— (1987), *The Economics of Bargaining*, Oxford: Basil Blackwell.

Brams, S (1977), 'Deception in 2 × 2 Games', *Journal of Peace Science*, **2**.

—— (1985), *Superpower Games*, New Haven, Conn.: Yale University Press.

—— and Hessel, M. (1982), 'Absorbing Outcomes in 2 × 2 Games', *Behavioral Science*, **27**.

—— and —— (1984), 'Threat Power in Sequential Games', *International Studies Quarterly*.

—— and Wittman, D. (1981), 'Non-myopic Equilibria in 2 × 2 Games', *Conflict Management, and Peace Science*, **6**.

Clausewitz, C von (1955), *De la guerre*, French translation, Paris: Minuit.

Guyer, M. and Hamburger, H. (1968), 'A Note on the Enumeration of All 2 × 2 Games', *General Systems*.

Harsanyi, J. (1967–8), 'Games with Incomplete Information played by "Bayesian players"', *Management Science*, **14**.

—— (1977), *Rational Behavior and Bargaining Equilibrium in Games and Social Institutions*, Cambridge, Mass.: MIT Press.

—— and Selten, R. (1972), 'A Generalized Nash Solution for Two-person Bargaining Games with Incomplete Information', *Management Science*, **18**.

—— (1988), *A General Theory of Equilibrium Selection in Games*, Cambridge, Mass.: MIT Press.

Kahneman, D. and Tversky, A. (1979), 'Prospect Theory: an Analysis of Decision under Risk', *Econometrica*, **47**.

Kohlberg, E. and Mertens, J.F. (1986), 'On the Strategic Stability of Equilibria', *Econometrica*, **54**.

Kreps, D. (1990), *Game Theory and Economic Modelling*, Oxford: Clarendon Press.

Lewis, D. (1961), *Convention: A Philosophical Study*, Cambridge, Mass.: Harvard University Press.

Osborne, M. and Rubinstein, A. (1990), *Bargaining Markets*, San Diego, Cal.: Academic Press.

Powell, R. (1990), *Nuclear Deterrence Theory*, New Haven, Conn.: Yale University Press.

Raiffa, H. (1982), *The Art and the Science of Negotiation*, Cambridge, Mass.: Belknap Press of Harvard University Press.
Rapoport, A., Guyer, M. and Gordon, D. (1966), *The 2 × 2 Games*, Ann Arbor, Mich.: University of Michigan Press.
Rubinstein, A. (1982), 'Perfect Equilibria in a Bargaining Model', *Econometrica*, **50**.
Rudnianski, M. (1986), 'Une approche de quelques problèmes stratégiques par la théorie des jeux', *L'Armement*, **4**.
—— (1991), 'Deterrence Typology and Nuclear Stability: a Game Theoretical Approach', in Avenhaus, Rudnianski and Karkar (eds), *Decision-Making Analytical Support and Crisis Management*, Heidelberg: Springer Verlag.
Schelling, T. (1960), *The Strategy of Conflict*, Cambridge, Mass.: Harvard University Press.
Schmidt, C. (1990), 'Dissuasion, rationalité et magasins à succursales mutiples', *Revue d'économie politique*, **5**.
—— (1991a), *Penser la guerre, penser l'économie*, Paris: Odile Jacob.
—— (1991b), 'Equilibria et accessibilité: repères pour une approche pragmatique des solutions d'équilibre de Nash', *Fundamenta scientiae*.
—— (1993a), 'Causality, Time and Economics', mimeo, Université de Paris Dauphine.
—— (1993b), 'L'*Homo bellicus* et le problème de la coordination économique', *Revue économique*, **3**.
Selten, R. (1978), 'Reexamination of the Perfectness Concept for Equilibrium Points in Extensive Games', *International Journal of Game Theory*, **4**.
—— (1978), 'The chainstore paradox', *Theory and Decisions*, **9**.
—— (1987), *Models of Strategic Rationality*, Dordrecht: Kluwer.
Shubik, M. (1984), *Game Theory in the Social Sciences*, Cambridge, Mass.: MIT Press.
Walliser, B. (1988), 'A Simplified Taxonomy of 2 × 2 games', *Theory and Decision*, **25**.

PART III

Dynamic Games and Information

6. Getting closer at different speeds: strategic interaction in widening European integration

Gerald Schneider*

INTRODUCTION

The purpose of the present study is to identify reasons for the current tendency of states to engage in political integration. The propensity to forgo national sovereignty in exchange for free-trade agreements and membership in regional IGOs contradicts Waltz's (1979, p. 105) statement that such developments are often talked about but seldom realized. On the contrary, the opposite holds true for the last decade of this century. Although regional co-operation frequently takes place, there has been hardly any progress in our understanding of such processes since the heyday of integration research in the 1950s and 1960s.

This theoretical stalemate is also manifest in studies about the enlargement ('widening') of regional integration. At present, no systematic explanation is available for the puzzling variation in the integration policies of otherwise comparable states: some want to integrate as quickly as possible while others are much more reluctant to sacrifice national sovereignty. However, the speed of nations is not an explicit topic in the classical approaches. In the perspective of neo-functionalism, it does not matter that governments pursue different interests (Haas, 1958, p. 524). Realist contributions (Moravcsik, 1991) are also of little help because they define state interests exogenously. Finally, studies focusing on political culture (Janssen, 1991) do not explain how the preferences of constituents should affect the positions of the negotiating governments.

Conceiving of integration as a principal-agent interaction between governments and their constituents, this chapter tries to shed some light on national strategies in the widening of regional integration. I develop micro-foundations of interstate co-operation for this purpose and define the preferences of parti-

* This chapter is a revised version of a paper presented at the 1992 Annual meeting of the American Political Science Association in Chicago and the Inaugural Pan-European Conference of the ECPR Standing Group on International Relations, Heidelberg. For comments, I am indebted to Mike Anthony, Lars-Erik Cederman, Cédric Dupont, Sieglinde Gstöhl, and Simon Hug. Support by the Swiss National Science Foundation is gratefully acknowledged (Grant no. 8210–030615).

cipating governments endogenously. The chapter particularly shows that domestic institutions and the dependence of governments on voter support influence the negotiation strategies. Governments have to take into account the interests of their principal who only approves of a beneficial treaty. Ratification procedures accordingly determine how much latitude a negotiator has. Since unpopular governments try to preserve sovereignty to survive domestically, the outcome of such a process furthermore depends on electoral considerations. Both kinds of constraints shape the integration process in an ambiguous way. While they yield more favourable outcomes for some governments, they also enhance the likelihood of bargaining inefficiencies.

To explore the linkage between international and domestic politics, I present models of the prenegotiations, the interstate bargaining and the ratification of an integration treaty. The prenegotiations are visualized as a Stackelberg–Cournot game. This model shows that governments can be treated as unitary actors even if there is both domestic and international disagreement about the level of integration. Nevertheless, the negotiation positions depend upon the preferences of the constituents in all participating countries. In the formal bargaining, informational incentives are of the utmost salience. If the uncertainty about domestic constraints is asymmetric, some unconstrained governments obtain better solutions by imitating the behaviour of constrained governments. Due to this uncertainty, negotiations might even fail although all actors prefer the enlargement of the integration process. Additionally, governments may also fall into an integration trap because they cannot figure out what the constituents really want. In the ratification game, finally, some governments attempt to hide how much sovereignty they have sold. With incomplete information, the pivotal constituent can commit the mistake of rejecting a beneficial treaty.

The chapter is arranged as follows: Section 1 summarizes the literature and selects the relevant levels of decision-making; section 2 outlines the basic conflict of interest in regional integration. The formal models are presented in section 3. The empirical illustrations in section 4 refer to the talks between the European Free Trade Association (EFTA) and the European Community (EC) about the European Economic Area (EEA). Section 5 summarizes the results and suggests future avenues of research.

1 REGIONAL INTEGRATION AS A MULTILEVEL GAME

1.1 The Impact of Other Decision-making Levels

Although regional integration currently progresses on various continents, our understanding of such processes does not coincide with this new momentum. A sign of this theoretical stalemate is the tendency in many current studies to borrow from the neofunctional (Haas, 1958, 1964) and the transactional traditions (Deutsch *et al.* 1957). While expressing varying degrees of scepticism towards their usefulness, some authors explicitly refer to the very broad categories of these theories (George, 1985, and Mutimer, 1989, for a neo-functional framework; Wallace, 1990, for a transactional set-up). As a result, these examinations reflect almost involuntarily the scientific discourse which prevailed in the 1950s and 1960s.

Regardless of the allusions to 'spillover effects' or other influential concepts, the classical approaches no longer seem apt to explain how political co-operation evolves at a regional base. This is to a large extent a result of the reluctance to accept intergovernmentalism (Scharpf, 1988) as the dominant decision-making mode. If supranational institutions, rather than governments, are the main actors, the divergence in the integration strategies of different states do not count. In the view of Haas (1958, p. 524), 'no government habitually seeks to hinder or advance integration'. To overcome the legacy of supranationalism, Moravcsik (1991, p. 688) recently proposed that 'the primary source of integration lies in the interests of the states themselves'. In other words, this kind of approach takes national preferences as exogenously given and resorts to a unitary-actor assumption. This is similar to examinations which focus on political culture as an explanatory concept. If the authors of such studies want to speak about integration policy they have to presume a direct link between voter attitudes and the negotiation style of governments (Janssen, 1991).

Owing to the necessity of a domestic vote on important integration issues this piece of research cannot solely rely on concepts such as the 'national' interest. In regional co-operation between two states at least four actors are crucial: two negotiators and two domestic principals, be it the cabinet, the parliament or the electorate. By disaggregating the preferences of states this chapter attempts to explore especially the impact of different decision-making levels on integration. Due to the importance of strategic considerations and the interactive nature of such an endeavour, I analyse regional integration from a rational-choice viewpoint. I assume that integration follows a rather strict temporal order. Such a process begins with the suggestion by one government to another to establish common political institutions. This proposal might lead to formal negotiations which are concluded by way of ratification. These approval procedures range from voting on the cabinet level to parliamentary decisions and referenda.

In sum, I assert integration to be a principal–agent situation at the domestic level. Yet, in contrast to less salient negotiations, governments do not act solely as agents of their constituents. This assumption is in accordance with an observation by Schelling (1960, p. 29). He described such negotiating agents as principals in their own right, with a preference order possibly deviating from that of its domestic constituents. As a result governments do not try to defend the 'national interest', but they are concerned about their own popularity and other individual matters. By asking for a 'better' treaty they can compensate for crises in other arenas of government activity. Due to this connection with other conflicts, integration is a multilevel interaction with the negotiations being the central decision-making environment.

1.2 New Approaches Towards an Old Topic

The interaction between various levels of decision-making has gained considerable importance in the fields of international relations and comparative politics. Among all interfaces, the relationship between international and domestic politics still attracts most attention. For a long time the controversy between realism and its liberal counterparts dominated studies on this subject. Due to a preoccupation with hierarchical considerations the discussions remained centred on the alleged superiority of either international or domestic sources of foreign policy (Almond, 1989).[1]

Instead of proffering a futile rank order, it seems more fruitful to investigate the factors under which both kinds of circumstance play a role. By analysing the interactions between the two levels in this way, Putnam (1988) offered a first rigorous treatment of the issue. In a formal development of these ideas, Iida (1991) shows how a government may benefit from asymmetric information. This study follows a similar approach and highlights the role of informational incentives in different periods of an integration process. In contrast to Iida's application of the Rubinstein (1982) model, the bargaining is, however, about discrete options and does not involve discounting.

Along with another attempt to put forward a general model of two-level games (Dupont, 1992), several indirect applications of this notion exist. In the domain of arms-control negotiations such models involve bargaining among allies (Brams, 1990) or between a government and its electorate (Morrow, 1991). Bueno de Mesquita and Lalman (1992) study the impact of domestic opposition on the beliefs and actions prior to war. Relying on an explicit principal–agent framework, Richards *et al.* (1992) contemplate the conditions under which governments engage in diversionary foreign policy behaviour. In comparative politics, studies on legislative committees (for example, Baron and Ferejohn, 1989) and on government formation (such as Laver and Shepsle, 1990) are similar to the idea of a nested interaction.

While the majority of these authors rely exclusively on the tools of non-cooperative game theory, the most acclaimed approach for building up a unifying approach draws upon Aumann's (1974) concept of correlated strategies. According to Tsebelis's (1990) notion of 'nested games' the potential for co-operation can be enhanced by allowing players to communicate. Another general idea is that the payoffs of the players in a principal arena vary according to the situation prevailing in other contexts. In other words, the general payoff attributed to an outcome could represent some combination of the payoffs in all relevant conflicts. However, it should be noted that this method hardly provides single equilibria. In technically sophisticated extensions (McGinnis and Williams, 1991), no clear threshold between conflict and co-operation exists. In a critique, Kreps (1989, p. 19) calls for the explicit inclusion of correlating devices and enforcement mechanisms. As every textbook describes integration as a process (for example, Groom and Heraclides, 1985, p. 174), it seems appropriate to rely on extensive form games

2 THE ENLARGEMENT OF AN INTEGRATION PROCESS AS A NESTED INTERACTION

In defining a national strategy towards regional integration, governments are confronted with a severe tradeoff. They must strike a balance between the loss of sovereignty and the prospects of welfare gains. At the same time, they are compelled to take other domestic conflicts into account. In a general manner, the following analysis illustrates in what respect popularity becomes important for the decision on the extent of integration. It furthermore shows how some leaders may use domestic resistance against a loss of sovereignty as a negotiation tool. I shall call this government the 'applicant'. In negotiations about the widening of an integration process, the applicant typically tries to preserve more sovereignty than the existing intergovernmental organization (IGO) would like to grant. Both sides perceive integration, however, to be better than the status quo. No bargaining would take place otherwise

To begin with the perspective of the constituent in the applicant country, I assume that the parliament or the electorate has the final say about the likely negotiation outcome. As the final decision will be taken with respect to some kind of majority rule, a pivotal constituent such as the median voter plays the central role. By comparing the costs of losing national sovereignty with the benefits of integration, this actor delimits how much flexibility a government has. Figure 6.1 shows how the median voter takes position with regard to political integration. The horizontal axis depicts how far the formal links between two entities can go. For simplicity, the extent of integration is an interval [0,1], with 0 representing full independence and 1 complete unification. The vertical axis measures the costs and benefits of political integration.

The curve *OC* displays the costs associated with different degrees of political integration. It seems reasonable to expect a rising slope. As a consequence, the more national sovereignty the applicant government decides to trade in, the more a voter has to give up. To mention but a few such losses, the individual sacrifice consists of giving up protectionist rents, rising information costs, a growing anonymity due to the enlargement of a decision-making system, and the shrinking importance in political matters of the own state. *OB* is the curve which indicates the benefits obtained from integration. Such benefits include the growth in personal income and in influence on the politics of other states.

The ideal point of the median voter I_M is where the difference between total costs and total benefits reaches its maximum. This is of course the point where the marginal total costs and the marginal total benefits are equal. As the pivotal constituent would approve of all outcomes promising higher benefits than costs, the intersection between *OC* and *OB* bounds the zone of agreement. Possible integration treaties lie within the range OU_M. Figure 6.1 indirectly shows the impact of different ratification procedures. By letting a qualified majority decide, the zone of agreement contracts.

Figure 6.1 The median voter and political integration

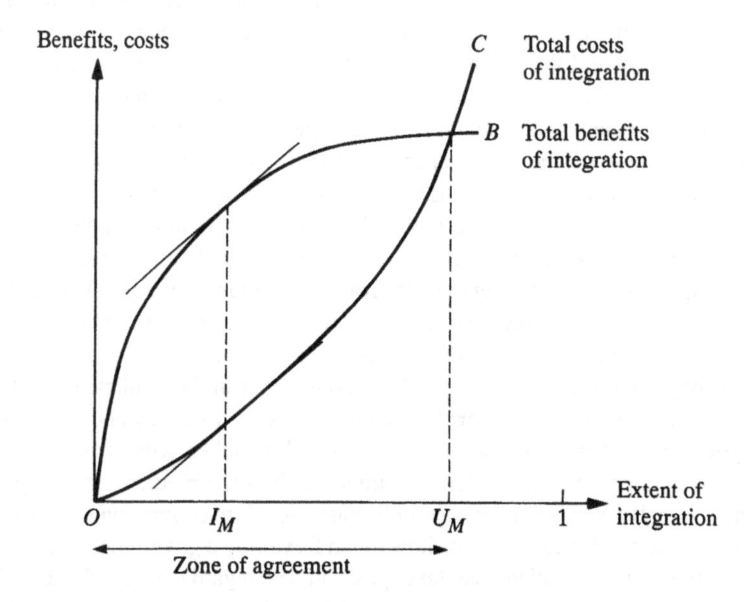

The following equations summarize the utility that voters C_{11} and C_{21} can derive from an integration attempt:

$$U^*_{c11(x)} = b_{11} - (x_1 - c_{11})^2 - r_{11}x_1$$

$$U^*_{c21(x)} = b_{21} - (x_1 - c_{21})^2 - r_{21}x_2$$

The ideal points of the applicant government and the IGO are g_1 and g_2 respectively; their constituents prefer most c_{1i} and c_{2j} ($i = 1, \ldots n; j = 1, \ldots m$). All actors have quadratic loss functions. Constituents can receive a maximal benefit $b_{1i,}$ and b_{2j}, from an integration treaty. They punish or reward a government upon evaluating how beneficial the outcome is. The scalars r_{1i} and r_{2j} represent the respective choices. They translate in the form $r_{1i}x_1$ and $r_{2j}x_2$ into rewards for the governments and voting costs for the constituents. Because of rising information costs I assume the voting costs to grow linearly with the level of integration. In other words, it becomes more and more difficult to receive adequate information about national integration policies the closer two political entities move towards unification and the less important the national level is.

The ratification constraint is the point where the utility of integration is zero for the pivotal constituent. Assuming negligible voting costs, this restriction is defined as follows for the principal of the first country:

$$x_1 = c_{11} + \sqrt{b_{11}}$$

Within their respective ratification limits leaders are basically free to choose exactly how intensive the established links with another entity should be. Although ideology may matter up to a certain point, domestic competition can prevent partisan integration policies. The negotiation position depends, in particular, on the support it enjoys for its general performance. In a period of high unemployment, for instance, its popularity may be exceedingly low. This creates an incentive to link the integration issue to other policies in order to improve domestic reputation. In the extreme, integration is embedded into a nested conflict in which executives try to maximize the support they can gain across all relevant policy arenas. Figure 6.2 indicates how this connection may affect the set of possible agreements.

OL displays the curve of votes an executive stands to lose when it opts for further integration. The curve *OG*, on the other hand, suggests how many votes it may gain from a specific policy. If an unpopular leader seeks additional votes in an attempt to stay in power, agreements are less likely to be reached. As the government in Figure 6.2 has, as a minimum, to seek a popularity gain *MG*, the zone of agreement becomes smaller. UG_L illustrates the lower and UG_U the upper level of the set of feasible agreements.

In formal terms, leaders can expect a maximum popularity reward $\Sigma r_{1i}x_1$ and $\Sigma r_{2j}x_2$ for a single integration treaty. If the treaty does not coincide with their ideologically preferred position, they suffer a quadratic loss $-(x_1 - g_1)^2$. This disutility is multiplied with a constant a_i which expresses to what extent the government depends upon its own popularity. The more sensitive a government is to changes in public support the more the deviation from the ideal point to the bargaining position matters. Accordingly, a_i is bigger than 1 for an unpopular government, and the range of this crucial parameter is between 0 and 1 for popular governments.

Figure 6.2 The position of an unpopular government with regard to integration

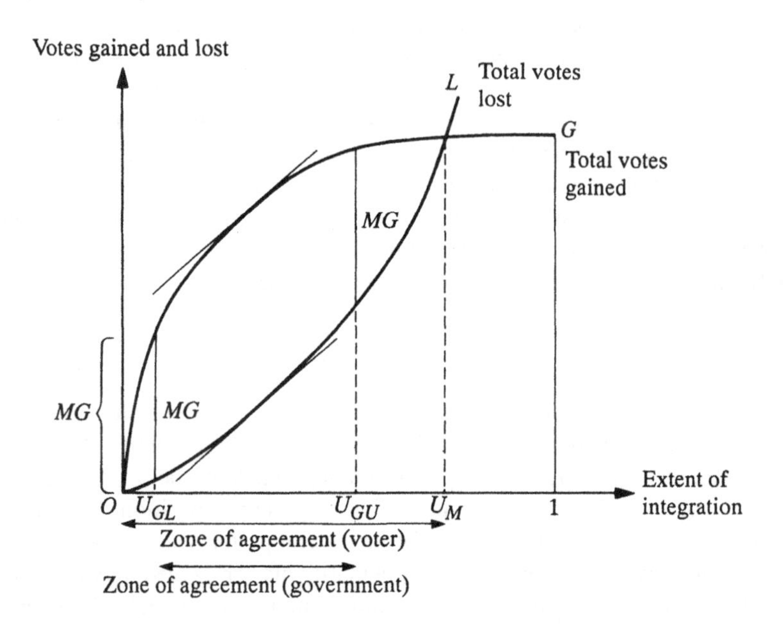

At the international level a strategic interaction unfolds because the two negotiation positions are not identical. The larger the distance between x_1 and x_2, the more cumbersome it is to agree upon an integration treaty. As a consequence, the popularity functions include the popularity loss $-(x_1 - x_2)^2$ and $-(x_2 - x_1)^2$. As governments have to respect the domestic politics of other countries to varying degrees, this disutility mirrors the power structure in the international system. The parameters w_1 and w_2 represent the weights which the two governments have to attribute to each other. As these exogenously given coefficients are relative measures of bargaining power, they add up to 1.

In sum, the popularity functions of the two governments accordingly look as follows:

$$U^*_{G1(x)} = \sum_{i=1}^{n} r_{1i}x_1 - a_1(x_1 - g_1)^2 - w_1(x_1 - x_2)^2$$

$$U^*_{G2(x)} = \sum_{j=1}^{m} r_{2j}x_2 - a_2(x_2 - g_2)^2 - w_2(x_2 - x_1)^2$$

Although the spatial analysis is helpful for developing the utility functions it is necessary to move beyond intuition. To obtain a more precise picture of the interaction between various decision-making levels I shall distinguish three different periods of integration negotiations: the prenegotiations, the formal bargaining and the ratification process.[2]

3 THREE PERIODS OF REGIONAL INTEGRATION

3.1 Defining the Bargaining Position

Because both governments and their constituents define how much their countries shall be formally linked with another political entity, regional integration can be represented as a nested two-stage game. This first becomes obvious in the prenegotiation period which I shall describe as a Stackelberg–Cournot game.[3] To assume the strict sequentiality of a Stackelberg game may be objectionable. However, integration negotiations are so salient that domestic actors try to influence the government strategy before the interstate negotiations commence.

Domestically, the interaction is a Stackelberg game in which the constituents move first to establish their position. This leads them to incorporate the executives' optimal reaction into their decision of whether they want to support the integration policy. Thus, they are the Stackelberg leaders, whereas the second-moving governments are the Stackelberg followers. On the international level, the interaction represents a Cournot game in which the governments choose their negotiation positions simultaneously. By incorporating the interests of the other side, the two executives have to take one another into account.

Despite the pressure from both the domestic and the international level, the preferences can still be consistently aggregated into negotiation positions for each country. Proposition 1 describes this result for the applicant government.

> *Proposition 1:* If the applicant government has to respect the interests of its constituents and to take into account the strategy of the regional IGO, its integration negotiation position represents the weighted average of the executives' bliss points in both entities.

The appendix includes the derivation of this result. The main result of the Stackelberg–Cournot prenegotiation game is that the lack of domestic consensus does not prevent governments from adopting a consistent policy. The model furthermore shows that governments still may be treated as unitary actors if they are exposed to pressure from other governments. The international linkage only forces them to come to terms with domestic as well as international considerations. The prenegotiation game also allows the definition of the negotiation positions endogenously. This is in contrast to the state-centric integration literature. Substantially, the possibility of an unambiguous negotiation position adds to the recent criticism of the bureaucratic politics model (Bendor and Hammond, 1992). In an implicit way, the 'third' model presented in the 'Essence of Decision' (Allison, 1971) is a reaction against rational choice and the unitary-actor assumption.

In other words, position-taking in both countries influences government strategies with regard to an upcoming international negotiation. The dependence on government support is ambiguous. The rewards which the applicant receives influence the willingness to integrate of the IGO. Because their zones of agreement are generally smaller, unpopular leaders ($a_1 > 1$) are less integration-minded than popular leaders. As the sensitivity parameters a_1 and a_2 grow, the size of the negotiation positions thus decreases. Finally, if a government is internationally omnipotent ($w_1 = 0$ or $w_2 = 0$), it has only to take into account its own domestic struggle.

This first model offers a general framework and highlights the mutual interdependence of negotiators. It furthermore shows that international unanimity ($x_1^* = x_2^*$) about the extent of integration is rare. In opposition to functionalist reasoning, national strategies therefore influence the integration processes, and bargaining is a necessary part of this kind of interstate interaction.

3.2 Exploiting and Enduring Domestic Constraints

The Negotiation Game models a situation where the two agents negotiate about two qualitatively different treaties. The first outcome (More Treaty) promises a higher level of integration than the less ambitious result (the Less Treaty). In the context of the enlargement of an IGO, the More Treaty could represent the 'entry conditions' of the IGO. The Less Treaty stands for a solution where the applicant would have to sacrifice less sovereignty. To move from the Less Treaty to the More Treaty involves the potential member in giving up a central symbol of its independence. The division into only two options reflects the fact that many political decisions are dichotomous, especially if electoral considerations are important. These domestic circumstances force politicians to choose between integration with or without an issue, such as a common currency.

In such a bargaining situation the IGO has to reckon with the power of the weak negotiator (Allan, 1984). If states are disaggregated into principals and agents, domestically constrained governments turn out to be the tough negotiators. To obtain more favourable outcomes, they can employ different kinds of threats (Schneider and Cederman, 1992). The Negotiation Game models the consequences of one of these manipulations. I assume that the applicant has private information about the preferences of its principal. To put it another way, the IGO does not know the ratification prospects of the other negotiator. The informational advantage of the applicant refers to the monopoly this player has in influencing its own public opinion. As in every signalling game, the presence of incomplete information leads to the distinction between different senders of a message. In Figure 6.3, two types of 'applicant' governments are presented. The principal of the strong applicant (player *SA*) never approves of the More Treaty whereas the constituents of its weak counterpart (*WA*) always ratifies such an agreement. The applicant prefers the Less Treaty to the More Treaty ($l_1 > m_1$) whereas the IGO opts for the opposite ($m_2 > l_2$).

Figure 6.3 Integration negotiations with incomplete information (ratification threat)

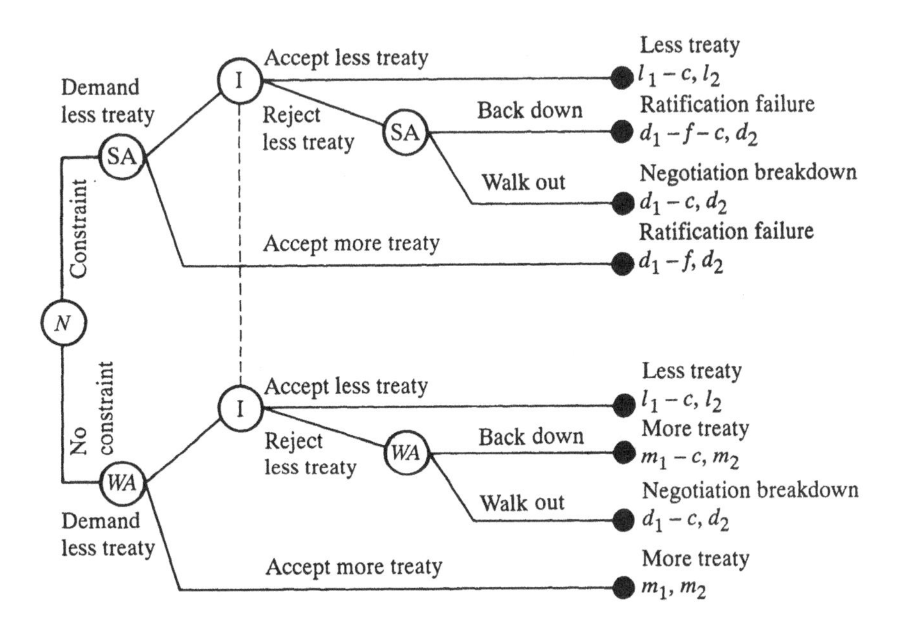

The game starts with a move by nature. The applicant has to decide whether to succumb to the More Treaty or to demand the Less Treaty. If it opts for the preferred Less Treaty, the IGO can in return insist on the More Treaty or accept the Less Treaty. The demand for the More Treaty forces the applicant either to walk out of the negotiations or to back down. In the event of a walkout the status quo prevails. The two negotiators can count on the deadlock payoffs d_1 and d_2 for such an outcome. Furthermore, the applicant has to reckon with different kinds of costs. First, it receives a penalty f if its principal refuses to support the More Treaty. Secondly, the commitment to the Less Treaty invokes signalling costs c. If a strong type subjects a More Treaty to ratification after demanding the Less Treaty, it obtains the payoff $d_1 - f - c$. For a weak type, the corresponding utility amounts to $m_1 - c$.

The Negotiation Game is based on the following assumptions:

Assumption 1: $d_1 - f - c < d_1 - f < d_1 - c < m_1 - c < m_1 < l_1 - c$
Assumption 2: $d_2 < l_2 < m_2$

Notice that all players prefer integration to a negotiation breakdown ($d_1 < m_1$ and $d_2 < l_2$). The analysis shows that integration negotiations under uncertainty are very different from bargaining under complete information. If the IGO knows the applicant's type, such negotiations are always efficient. Neither of the applicants walks out of the negotiations and the weak applicant never attempts to bluff. On the contrary, this player immediately accepts the More Treaty, anticipating that the IGO would reject the demand for a Less Treaty. The strong applicant, by contrast, obtains the Less Treaty. Proposition 2 summarizes this result:

> *Proposition* 2: Under complete information, the Negotiation Game has two outcomes:
> EQUILIBRIUM I: If the IGO is certain to encounter the weak applicant, the negotiations result in the More Treaty.
> EQUILIBRIUM II: If the IGO is certain to encounter the strong applicant, the negotiations result in the Less Treaty.

Enlargement negotiations are, however, much more ambiguous as soon as the assumption of complete information is abandoned. If there is uncertainty about the applicant's constituent, the integration process may lead to suboptimal outcomes. Proposition 3 summarizes this case. A sketch of the proof for the two propositions can be found in the appendix.

> *Proposition 3*: Under incomplete information, the Negotiation Game has two major outcomes:
> EQUILIBRIUM 1: Up to a certain threshold belief $p_0 = (m_2 - l_2)/(m_2 - d_2)$, the strong applicant always demands the Less Treaty. As the IGO sometimes rejects this demand, both walkouts and Less Treaties are possible outcomes. The weak applicant randomizes the first move. If this player demands the Less Treaty, the IGO also adopts a mixed strategy. This leads to the Less Treaty or to the backing down outcome.

EQUILIBRIUM 2: Above the threshold belief p_0 both the weak and the strong applicant always asks for the Less Treaty which is, subsequently, accepted by the IGO.

The outcome of integration negotiations under uncertainty depends on whether the IGO strongly believes that it will encounter a constrained applicant. The applicant's demands are always successful above this borderline belief. The suboptimal outcome of a negotiation failure only becomes possible below the crucial belief. The walkout from the negotiations happens when the IGO erroneously believes it faces a weak applicant. In this situation the weak applicant can successfully bluff, although some of these attempts might be called.

Although their reasons differ, both types have an incentive to negotiate. The strong applicant derives its motivation to ask for a concession from the penalty for the ratification failure. In the absence of carrot-and-stick measures by his principal, this type of negotiator would not mind subjecting the More Treaty to ratification and experiencing a defeat. By contrast, the likelihood that its weak counterpart immediately accepts the More Treaty only depends on the deadlock payoff d_2. If the IGO enhances this payoff, the More Treaty becomes more likely; or, to put it another way, the power of the IGO grows as long as the negotiations become less salient.

3.3 Circumventing Constraints During the Ratification

However important private information may be, bilateral uncertainty is a very common trait of integration negotiations. Negotiating agents do not necessarily know whether a principal may ratify the envisioned treaty. Due to the veto power of the constituents some governments accordingly fall into a decision-making trap in which both ratification failure and a negotiation breakdown are very likely outcomes. The higher the uncertainty about the principal the more reluctant the strategy of the government (Lax and Sebenius, 1991). Although a constrained government is strong at the international level, rigid ratification procedures may turn this advantage into a disadvantage in the domestic setting.

In contrast to the prenegotiations, the executives at least have the privilege of moving first in the ratification process. When approval is uncertain, they have an incentive to ensure a success. In this last model[4] (Figure 6.4), the government possesses private information about the outcome of integration negotiations. In accordance with the preceding situation, bargaining can either result in a More or a Less Treaty. A government is strong (player *S*) domestically if it brings home a Less Treaty. Its weak counterpart (player *W*), however, accepted a More Treaty at the international level. The constituent with the decisive vote (player *C*) would like to reject a More Treaty, but accepts a Less Treaty. It should be noted that this model is naturally only relevant for cases where ratification matters.

Figure 6.4 The Ratification Game

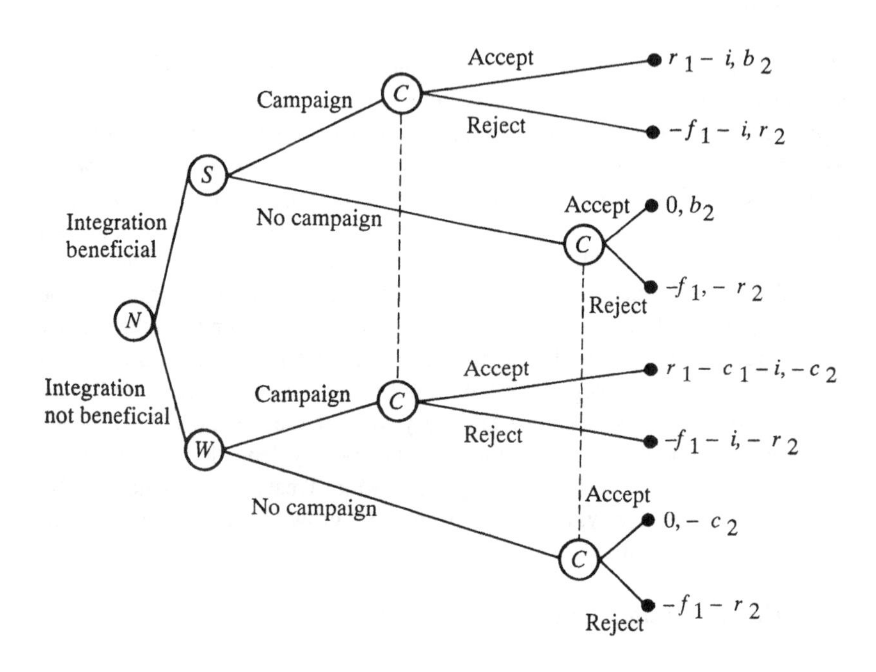

The game commences with a move by nature which determines the level of the treaty. Subsequently, the government has to decide whether it wants to convince the pivotal constituent that the treaty is beneficial. Regardless of the decision by the government, this crucial member of the electorate has the last word. If the information campaign succeeds and the treaty passes to ratification, the government obtains a popularity bonus r_1. A successful ratification without campaign, by contrast, does not have any effect on government popularity. Since governments cannot count on any change in their popularity, the payoff for an effortless success is zero. In the case of a ratification failure, governments are punished with a penalty f_1. To engage in an information campaign invokes a cost i additionally. A weak type has moreover to take into account that the decisive constituent will withdraw its long-term support when it realizes that the treaty was not advantageous. This second cost is denoted as c_1.

To mistakenly approve of the More Treaty imposes a cost c_2 on the constituent. When accepting a beneficial treaty, the constituent can count on the benefit b_2. The voting costs are assumed to be r_2. The payoff orders are thus defined as follows:

Assumption 3: $-f_1 - i < -f_1 < 0 < r_1 - c_1 - i < r_1 - i$
Assumption 4: $-c_2 < -r_2 < b_2$

The inequality $0 < r_1 - c_1 - i_1$ denotes that the weak type has an incentive to send a message. Yet this player tries to reach this outcome only if the constituent is uncertain about the outcome of the international negotiations. Under complete information there would again be no inefficiency in such a game. This situation is summarized in the following proposition:

> *Proposition 4*: Under complete information, the Ratification Game has two outcomes:
> EQUILIBRIUM I: If the government can present a beneficial treaty, it will engage in an information campaign and the constituent will accept the treaty.
> EQUILIBRIUM II: If the government cannot present a beneficial treaty, it will refrain from an information campaign and the constituent will reject the treaty.

Since the constituent knows the identity of the governments under these circumstances the strong type would send a message and the Less Treaty would pass the ratification process. Weak types, however, abstain from any promotion, and the ratification fails. This kind of applicant can only mimic the behaviour of its strong counterpart in the presence of asymmetric information. The outcome largely depends on the beliefs of the pivotal constituent of facing a strong applicant after it had observed a campaign. This situation is described in the fifth proposition:

> *Proposition 5*: Under incomplete information, the Ratification Game falls into two basic cases:
> EQUILIBRIUM 1: Up to the threshold belief $p^* = (c_2 - r_2)/(c_2 + b_2)$, the strong government always starts a campaign. The pivotal voter randomizes its strategy in order to make the weak government indifferent. This unsuccessful negotiator in return equivocates between campaigning and refusing to start a campaign. The constituent rejects the treaty if no message was sent.
> EQUILIBRIUM 2: The government always starts an information campaign and the voter always accepts it above the threshold belief p^*.

The model establishes that there is always an incentive to start a campaign for a strong applicant. It indicates that the uncertainty about the level of the treaty may induce suboptimal results. Hence, a government might be successful in pretending that it achieved a Less Treaty although it had to accept a More Treaty. On the other hand, pivotal constituents might mistakenly reject the strong applicant's Less Treaty, suspecting that the information campaign was not sincere. Comparative statics reveal that the likelihood of such an outcome increases with growing rewards for the successful information campaign (r_1), decreasing signalling costs i and a growing size of the penalty f_1 for a ratification failure. The probability of a successful bluff increases when the weak government commits itself strongly to the information campaign. The opposite effect occurs if this type is about to lose considerable electoral support after the pivotal constituents find out that they voted in favour of the More Treaty instead of the Less Treaty. The occurrence of bluffs thus depends very much on whether governments can be held accountable for what they said during a ratification campaign. It is obvious that the size of such costs grows the more competitive a political system is.

Considering the crucial impact of such possible punishments, the dependence of negotiations on their domestic setting has thus been again the general feature of this last integration period. I demonstrate the empirical relevance of the different models by describing the negotiations between the European Free Trade Association (EFTA) and the European Community (EC) about the creation of an European Economic Area (EEA). Analytically, these talks represented a three-level game, consisting of the interactions between the EC and the EFTA, the intraorganizational coalition-building and the domestic negotiations in each state. Substantively, the goal of the EEA is to enlarge the scope of European integration and to extend the Internal Market Programme to the EFTA.

4 THE EEA NEGOTIATIONS AS A THREE-LEVEL GAME

1.4 Reactions to the Internal Market Programme

It was Austria which reacted most fiercely to the Internal Market Programme of the EC. A coalition of business forces particularly influenced the junior partner in the government, the conservative ÖVP, to request full membership of the EC (Luif, 1990, p. 184). After the dominant Socialist Party had followed suit the National Parliament voted in favour of such a step. The government applied for full EC membership on 17 July 1989. At this time such a move was out of the question for the other EFTA members.[5] They still hoped to be able to continue their pick-and-choose strategy and to strike separate deals with the EC. Bilaterally they were already closely tied to the Community. By 1989 the links of Switzerland alone amounted to around 150 separate accords.

This approach remained the main channel for integration despite some efforts to establish direct negotiations between the EFTA and the EC. At a joint summit in Luxembourg in 1984, the two organizations had solemnly declared their intention to create a 'dynamic European Economic Space'. Notwithstanding this proclamation, the supranational talks took off only very reluctantly. Almost half a decade passed before the idea achieved a prominent place on the integration agenda. In this new situation, the label of the project soon changed from 'European Economic Space' to 'European Economic Area', primarily for linguistic reasons.

4.2 Defining Integration Strategies at the National Level

A move by a single player turned the issue of how 'to incorporate other nations into the Internal Market Programme' into a three-level game. By asking for direct negotiations between the EC and the EFTA on this account, Jacques Delors (1989), the President of the EC Commission, reactivated the idea of an EEA in January 1989. By calling for a more structured partnership, he thereby temporarily succeeded in protecting the deepening of EC integration from the adhesion of neutral countries such as Austria. Simultaneously, Delors drove all other EFTA governments into a corner. They were now forced to co-ordinate their policies and to speak with one voice.

After Delors had unilaterally defined the bargaining rules, the governments of the EFTA member states had to circumscribe the extent to which their countries were prepared to integrate. Press leaks soon revealed which governments would enter the upcoming talks as hard-liners. A quantitative indicator for the negotiation positions is the number of exceptions (derogations) which the EFTA members tried to obtain from the set of relevant Community rules, the so-called *acquis communautaire*.[6] According to a British source (Select Committee, 1990), the EFTA countries aspired to various permanent exemptions: Switzerland (4), Iceland (3), Austria (2), Sweden (2), Finland (2) and Norway (2). For example, the governments tried to limit foreign ownership of land and to uphold high standards for health and environmental protection. Nevertheless, the Commission categorically refused any permanent exceptions to the *acquis* in May 1990. Notwithstanding this blow, the hard-liners gave in only so far as they were urged to. At a meeting in spring 1991, the EC refused to accept the following number of temporary derogations: Switzerland (21), Iceland (9), Norway (6), Austria (5), Finland (4) and Sweden (2).[7]

These figures confirm that Switzerland and Iceland were indeed the countries which tried the hardest to preserve their sovereignty. Switzerland tried to compensate for losses in this realm by insisting on a role in the shaping of Community decisions around the EEA. Sweden, Finland and Austria, on the other hand, were much less reluctant to accept EC rules. As these states were the first to opt for EC membership the formation of the EEA institutions did not much matter.

It is still puzzling that there was a greater convergence of interests between the governments of Sweden and Austria than between the administrations of neighbouring countries like Switzerland and Austria. As the evaluation of the EC and the EEA does not deviate very much in public opinion across the EFTA member states, cultural factors account only marginally for this divergence. As suggested in the models, the prospect of possible gains along with electoral and institutional constraints appear to be dominant in the negotiation positions. A nation desperately in need of the blessings of deregulation would be more keen to reach a liberalizing integration agreement than a country with a strong economy. In other words, a crisis-ridden country is more likely to accept a considerable loss of sovereignty. Among the political factors, popularity and ratification constraints are of the utmost importance. Table 6.1 summarizes the ratification requirements and the state of the economy during the negotiations (1989–91).

Table 6.1 Determinants of the negotiation positions in the EEA talks: GNP growth, inflation, and unemployment between 1989 and 1991, and ratification procedures in the EFTA member states (minus Liechtenstein)

	GNP Growth[a]	Inflation[a]	Unemployment[a]	Ratification procedures[b]
Austria	3.8 (2.8)	2.8 (3.2)	3.3 (3.4)	Parliament/QM
Finland	0.2 (–5.2)	5.4 (4.1)	4.9 (7.7)	Parliament/QM
Iceland	–0.3 (0.3)	13.8 (8.2)	1.5 (1.6)	Parliament/SM
Norway	2.1 (4.1)	4.1 (1.5)	5.1 (5.3)	Parliament/QM
Sweden	0.4 (–1.2)	8.1 (7.0)	1.9 (2.7)	Parliament/SM
Switzerland	1.8 (–0.2)	5.3 (6.2)	0.8 (1.2)	Referendum/SM[c]

Notes
[a]OECD Economic Outlook December 1991. The numbers in parentheses refer to 1991 only.
[b]SM = Simple Majority required, QM = Qualified Majority.
[c]Simple majority of both the voters and the states.

On the economic side, no EFTA country performed exceptionally well in the period under scrutiny. Yet the countries hit most ferociously by the recession of the early 1990s, Sweden and Finland, also pursued a moderate course during the negotiations. Furthermore, they were the first countries to apply for EC membership after Austria, which had experienced economic difficulties in the mid-1980s.

On the institutional side, only Switzerland and Liechtenstein[8] were required to hold referendums about the EEA treaty. Swiss direct democracy provides territorial minorities with a veto power because the ratification requires a majority of both the voters and the cantons (states). As the smaller cantons are less export-orientated and urbanized than the densely populated centres the ratification procedures benefit the protectionist forces. In Norway, the EEA had to find a qualified majority, but only as far as the *Storting* (national parliament) is concerned, where 75 per cent of the members of parliament must approve a treaty.

Norway provided the clearest example of the importance of popularity constraints and the connection of integration with domestic issues. According to a journalistic account, the Norwegian negotiation position eventually stiffened after the ruling Labour party experienced a major setback in local elections: 'Its new loss of popularity will further complicate Norway's problems (notably the concessions asked of it on fish)', commented *The Economist* (14 September 1991).

As the government and the major opposition parties in Austria and Finland were always in favour of the EEA agreement, the ratification constraint of two-thirds majorities in the national parliaments never represented major obstacles. Similar to the Swedish case, the Icelandic approval procedure requires only a simple majority. However, protectionist interests tend to be overrepresented in the parliament. Fifty-five per cent of the seats are given to regions in which fishing is the main economic activity, although these regions make up only 38 per cent of the population.

Because the EEA requires the ratification of all participating states and the European Parliament, domestic concerns also influenced the strategies of the EC member states. Hard-liners could profit from the unanimity rule that was a prerequisite for the final decision about the EEA. This enabled in particular the Southern members to ask for concessions with regard to fishing rights and the EFTA contributions to the structural funds. The importance of the veto threat even from a minor state became overt through the Greek insistence on obtaining a sharp increase in the license fees for trucks passing through Austria. This demand ran counter to attempts to protect the Alpine environment from excessive lorry transportation. On the eve of the final decision about the EEA, the EC could block the hampering claim by a majoritarian decision in a separate transit treaty with Austria and Switzerland. However, such a manoeuvre ceased to be an option when the EC was to make a decision on the whole EEA treaty the next day. By threatening to veto the EEA, Greece managed to get very close to its initial request.

4.3 Intraorganizational Threats and Counterthreats

Throughout the negotiations, the governments of the EFTA member states remained the central actors. Nevertheless, as they were forced to speak with one voice, they also had to reckon with intraorganizational dynamics as a second relevant level of decision-making. In this respect the EFTA had to reconcile the interests between obstinate and conciliatory governments. Owing to the unanimity requirement, the power was on the side of those who favoured keeping as much sovereignty as possible, even to the detriment of other EFTA states. By hinting at their domestic hurdles they also enhanced the risk of a negotiation breakdown. This intraorganizational exploitation of domestically less-constrained governments was, however, far from complete. Integrationist governments possessed a forceful counterstrategy as they could threaten to strike a separate deal with the EC. Accordingly, Austria, Sweden and Finland undermined the firm positions of other EFTA members by pursuing their EC membership applications in parallel.

Leaking information about the nationalistic integration course of Switzerland was among the attempts to isolate this laggard. Scandinavian officials employed this tool at the beginning of the negotiations and even considered moving ahead without the Swiss. This manoeuvre revealed the tension between partial and full integration which is a common feature of IGOs (Schneider and Cederman, 1992).

4.4 Different Bargaining Leverage and Discount Rates

There is still no full explanation for the decision by the EC to intensify direct negotiations with the EFTA. Delors justified his initiative of January 1989 as a wish to negotiate in a more efficient way. He might have counted on the possibility that the negotiating power of the EFTA would decrease through internal fighting. These factors were probably strong enough to compensate for the loss of bargaining power compared to a situation where the EC would have to negotiate with each individual country.

Originally, the President of the EC Commission promised the EFTA countries 'a new, more structured partnership with common decision-making and administrative institutions' (Delors, 1989). One year later he officially retreated from this offer. In Delors's view, the 'osmosis' between the two organizations 'must stop short of joint decision making, which would imply Community membership' (Delors, 1990, p. 9). In the end, the EFTA was indeed forced to leave practically all decision-making power to the EC. This became even more overt after an unforeseen objection by the EC's European Court of Justice which ruled compulsory a revision of the treaty. Accordingly, the EFTA had to acquiesce to the demand that all disputes should be subject to adjudication by the EC and not solved by a joint legal body.

At the level of the individual countries, the EC partially respected the wishes of the EFTA governments and granted a number of special transition periods: Switzerland (25), Iceland (20), Norway (19), Austria (14), Sweden (13), Finland (11). Switzerland obtained some temporary derogations in highly sensitive areas, such as foreign immigration and the recognition of diplomas. The Nordic countries in part achieved an extension on the limitations of the free flow of capital. Additionally, Iceland and Norway were able to obtain permanent permission to keep out foreign investment from their fishing industries. The EC furthermore allowed Switzerland to continue with its weight limit of 28 tons on lorries passing through the country.

Despite this partial success, the treaty failed to live up to the initial expectations of the more nationalistic members of the EFTA. Instead of being treated like 'equals', they largely had to accept the conditions set forth by the much more powerful EC. The EEA thus stands in great contrast to the free-trade agreements which the EFTA countries and the EC concluded in the early 1970s. It should be obvious that the divergence in bargaining power principally accounts for this imbalance in the early 1990s. While the EC gained strength during this period of bilateralism, the EFTA remained the organization of the rich European outsiders.

4.5 Trying to Avoid Referendums

Switzerland was the only larger EFTA country which had real problems in ratifying the agreement. As a consequence, its government postponed a clear commitment to the integration process, hoping for a major shift in public opinion. As it turned out, this change did not happen. On 6 December 1992, a narrow majority of the voters and a clear majority of the cantons rejected the agreement, driving Switzerland into international isolation and a domestic crisis. Even the voters of Liechtenstein accepted the agreement a week later. The popular vote in Switzerland necessitated renegotiations among the remaining EEA participants. The passive stance of the Swiss government also indicated that the penalty for ratification failure is small in a consociational democracy. Nobody was forced to retire after the political establishment had experienced this major defeat.

In the other EFTA member states, governments had fewer problems in securing a success. The Austrian parliament rejected an initiative by the Green Party to subject EEA membership to a referendum. In Iceland, the question of holding a referendum was, for a certain period, more salient as the major opposition parties (Progressive Party, People's Alliance and Women's Alliance) declared themselves in favour of a ballot. The two major Norwegian parties, the Labour Party and the Conservative Party, rejected similar demands by some part of the opposition (Party of Progress, Socialist Left Party and Christian Democrats). However, none of these governments faced major problems in the ratification process.

The ratification phase completes the circle of a process which started and ended in the domestic setting. On the national level, governments were compelled to incorporate the opinions of their decisive constituent. At the intraorganizational level, the conciliatory EFTA governments could, nevertheless, undermine the veto power of the less integration-minded governments by a counterthreat to apply for EC membership. Finally, an imbalance in bargaining leverage and in possible benefits stemming from a treaty forced the EFTA to accept a very unequal treaty.

5 CONCLUSION

Relying on a formal framework, this study has analysed the different speeds of nations in the enlargement of an integration process. It has examined the interaction between different decision-making levels in the creation and enlargement of a regional IGO. Three formal models show how domestic factors influence integration strategies. This perspective is opposed to traditional approaches which focus on the overall integration process and mainly invoke system-level or functional explanations. By offering micro-foundations of interstate co-operation, this chapter also contrasts with state-centric theories and cultural explanations.

More specifically, this chapter highlighted the importance of the relationship between domestic constituents (principals) and their governments (agents). If a government lacks popular support, it tries to compensate during the integration negotiations. This inhibits its zone of agreement internationally. The principal may execute considerable influence on the government through carrot-and-stick measures. However, integration processes can be inefficient in two respects. First, negotiations can fail although all participating governments wish to strengthen the ties with another government. Secondly, the pivotal constituent can erroneously reject a beneficial treaty. Such suboptimal outcomes may arise under the condition that one actor possesses private information.

I described the EEA negotiations as a paradigmatic integration case in which governments have to balance interdependence and national sovereignty. As the evidence for strategic interaction in this setting is, up to now, only anecdotal, subsequent empirical applications will deal with the interface between domestic and international politics in a more rigorous way. This first step has, on the other hand, already pointed out that it is high time to move the research on political integration beyond the agenda set up in the 1950s and 1960s.

APPENDIX

A.1 Prenegotiation game

This first part of the appendix contains the solution for the Stackelberg–Cournot game of section 3.1. In this respect, I have to establish that the prenegotiation positions are influenced by the preferences of the constituents in both countries. Proposition 1 states that the positions x_1^* and x_2^* represent the weighted average of the executive bliss points. The popularity functions of the governments are as follows:

$$U^*_{G1(x)} = \sum_{i=1}^{n} r_{1i}x_1 - a_1(x_1 - g_1)^2 - w_1(x_1 - x_2)^2$$

$$U^*_{G2(x)} = \sum_{j=1}^{m} r_{2j}x_2 - a_2(x_2 - g_2)^2 - w_2(x_2 - x_1)^2$$

As the leaders move domestically second, their equilibrium choices (x_1* and x_2*) can be easily derived from these functions:

$$x_1^* = (g_1 + w_1x_2)/(a_1 + w_1) + \Sigma r_{1i}/2(a_1 + w_1)$$

$$x_2^* = (g_2 + w_2x_1)/(a_2 + w_2) + \Sigma r_{2j}/2(a_2 + w_2)$$

The constituents influence the negotiation position through their reward-and-punishment system. They can influence the leader by their choice of the scalar *r* which is included as a linear voting cost in their utility functions.

$$U^*_{C11(x)} = b_{11} - (x_1 - c_{11})^2 - r_{11}x_1$$

$$U^*_{C21(x)} = b_{21} - (x_2 - c_{21})^2 - r_{21}x_2$$

The principals can anticipate how the executives change the positions x_1 and x_2 with respect to the size of r_{1i}. These conjectural variations amount to $\phi_1 = \frac{1}{2}(a_1 + w_1)$ and $\phi_2 = \frac{1}{2}(a_2 + w_2)$ respectively. The domestic constituents incorporate them in their calculations. The first-order condition for the utility function of the *i*th constituent in the first country is as follows:

$$0 = -2(x_1 - c_{1i})\phi_1 - r_{1i}\phi_1 - x_1 + g_1.$$

Summing up the decisions with respect to r_{1i}, the government of the first country can count on the following reward for its integration policy:

$$\Sigma r_{1i} = 2[-nx_1 + \Sigma c_{1i} - (a_1 + w_1)nx + (a_1 + w_1)ng_1].$$

The government incorporates these rewards into its negotiation position x_1* for the integration talks:

$$x_1^* = [g_1\{1 + n(a_1 + w_1)\} + \Sigma c_{1i} + w_1x_2]/[a_1 + w_1 + n(1 + a_1 + w_1)].$$

By analogy, x_2* can be found which is expressed entirely in terms of w_1 instead of w_2

$$x_2^* = w^*[g_2\{1 + m(1 + a_2 - w_1)\} + \Sigma c_{2j} + x_1 - w_1 x_1].$$

In this last equation, w* stands for the following weighting factor:

$$w^* = 1/[1 + a_2 - w_1 + m(2 + a_2 - w_1)].$$

Substituting x_2* and some manipulation leads to the negotiation position x_1* of the first country:

$$x_1^* = [g_1\{1 + n(a_1 + w_1)\} + \Sigma c_{1i} + w_1 w^* (g_2\{1 + a_2 - w_1\}) + \Sigma r_{2j})]/$$
$$[a_1 + w_1 + n(1 + a_1 + w_1) - w_1 w^* (1 - w_1)].$$

As stated in the proposition, the weighted executive bliss points thus completely represent the negotiation positions. Or to put it differently, governments can be treated as unitary actors even when confronted with both domestic and international competition. If a government is omnipotent, it need not care about the international aspect. Hence, its negotiation position can be expressed in a simplifyng equation:

$$x_1^* = [g_1\{1 + n(a_1 + w_1)\} + \Sigma c_{1i}]/[a_1 + w_1 + n(1 + a_1 + w_1) + n].$$

A.2 Negotiation Game with Ratification Threat

Here I sketch the proofs for Propositions 2 and 3. Complete proofs for an almost equivalent game can be found in Schneider and Cederman (1992). I shall refer to the strong applicant as player 1, to its weak counterpart as player 1' and the IGO as player 2. The two subgame perfect equilibria (SPE) described in Proposition 2 can be derived through backward induction. If the IGO is certain to encounter the weak applicant (Equilibrium I), U_2 (Demand *M*) = $m_2 > U_2$ (Accept *L*) = l_2. Anticipating this, player 1' chooses the More Treaty: $U_{1'}$ (Accept *M*) = $m_1 > U_{1'}$ (Demand *L*) = $m_1 - c$. In Equilibrium II, U_2(Accept *L*) = $l_2 > U_2$ (Demand *M*) = d_2. The strong applicant profits from this opportunity by calculating U_1 (Accept *M*) = $m_1 < U_1$ (Demand *L*) = $l_1 - c$.

The sequential equilibria of Proposition 3 can be described by the behaviour strategies *s*, *t*, *q* and the organization's posterior belief *p*'. The parameter *s* stands for the probability of player 1 choosing Demand *L*. Since U_1 (Exit) = $d_1 - c >$ U_1 (Accept *M*) = $d_1 - f$, $r = 1$ by assumption. The likelihood for player 1' to formulate a demand is *t*. The IGO asks for the More Treaty with probability *q*.

These definitions can be used for the derivation of the equilibria 1, la and 2. It should be noted that the likelihood of equilibrium la is very small.

EQUILIBRIUM 1: If player 1' randomizes its moves, $U_{1'}$ (Demand L) = $q(m_1 - c) + (1 - q)(l_1 - c) = U_{1'}$ (Accept M) = m_1, leading to $q = (m_1 + c - l_1)/(m_1 - l_1)$. As 2 always mixes its strategy, U_2 (Demand M) = $p'd_2 + (1 - p')\, m_2 = U_2$(Accept L) = l_2. This last equation can be used to establish the threshold belief $p' = p_0 = (m_2 - l_2)/(m_2 - d_2)$. Using Bayes rule ($p' = p/(p + (1 - p)t)$) we can finally derive that $t = (p/(1 - p))(l_2 - d_2)/(m_2 - l_2)$. This first equilibrium is partly pooling.
EQUILIBRIUM 1a: Given $s = 1$ and assuming $t = 1$, the IGO cannot distinguish the applicants. As a consequence, player 2 cannot update its beliefs ($p' = p$). If 2 still mixes its response, the equation U_2 (Demand M) = $p'd_2 + (1 - p')\, m_2 = U_2$ (Accept L) = l_2 still holds, leading to $p' = p_0 = p$. This knife-edge equilibrium is pooling.
EQUILIBRIUM 2: A second pooling equilibrium can be obtained if $t = 1$ and $q = 0$. This means that player 2 always gives in to a Demand L. U_2 (Accept L) > U_2 (Demand M) implies that $p_0 < p = p' < 1$.

A.3 Ratification Game

The strong applicant is player 1, its weak counterpart player 1' and the pivotal constituent is player 2. Equilibria can be described by the behaviour strategies s, t, q' and q'' and the posterior beliefs p' and p''. Player 1 engages with probability s into an information campaign, player 1' with probability t. After an information campaign, player 2 accepts the treaty with probability q'. In the absence of a message, the likelihood for the equivalent step is q''. Player 2 attributes the belief p' to facing player 1 after receiving a message, and p'' is the belief if no message was sent. These notations appear in proposition 4 which describes the complete information case.

> *Proposition 4*: Under complete information, the Ratification Game has two Subgame Perfect Equilibria (SPE):
> EQUILIBRIUM I: If the government can present a beneficial treaty, it will engage in an information campaign and the constituent will accept the treaty ($s = 1, q' = 1, q'' = 1, p' = 1, p'' = 1$).
> EQUILIBRIUM II: If the government cannot present a beneficial treaty, it will refrain from an information campaign and the constituent will reject the treaty ($t = 0, q' = 0, q'' = 0, p' = 0, p'' = 0$).

PROOF: As in Proposition 2, the concept of backward induction is used to derive the SPEs. In equilibrium I, player 2 is certain to encounter a strong government. Hence, its posterior beliefs p' and p'' are 1. To accept is thus a dominant strategy, with U_2 (accept/Campaign) $= b_2 > U_2$ (reject/Campaign) $= -r_2$ and U_2 (accept/No Campaign) $= b_2 > U_2$ (accept/No Campaign) $= -r_2$. Player 1 anticipates these choices. Because U_1 (Campaign) $= b_1 > U_1$ (No Campaign) $= 0$ by definition, this player starts an information campaign.

In equilibrium II, by contrast, player 2 is certain to encounter a weak government. The posterior beliefs are accordingly both 0. By assumption, rejection of the treaty must then always be a dominating strategy, regardless of the preceding action by the government: U_2 (reject/Campaign) $= -r_2 > U_2$ (accept/Campaign) $= -c_2$ and U_2 (reject/No Campaign) $= -r_2 > U_2$ (accept/No Campaign) $= -c_2$. To minimize the losses, player 1' in return does not start an information campaign: $U_{1'}$ (No Campaign) $= -f_1 > U_{1'}$ (Campaign) $= -f_1 - i$. QED

Assuming incomplete information, it is useful to establish first the relationship between s and t. In accordance with assumption 3, the likelihood of 1 choosing s is at least as big as the probability t. Given the preference order, the strong government is more likely to commit itself to an information campaign than its weak counterpart. I continue by testing whether the remaining possible combinations of s and t are part of $D-1$ equilibria (Banks and Sobel, 1987). The reference to an equilibrium refinement is necessary because the structure of the game allows for out-of-equilibrium beliefs.

Case 1: $s = t = 0$

As the upper information set is not reached in this case, I cannot use Bayes' rule to derive p'. There are beliefs such as $p' = 0$ which support this case as a sequential equilibrium. The $D-1$ criterion however allows this out-of-equilibrium belief to be discarded and, in consequence, this candidate equilibrium. According to the $D-1$ refinement, I have to check whether the strong applicant is more likely to defect from this equilibrium path. The weak government would thus only weakly prefer to engage in a campaign: $U_{1'}$ (No campaign) $= (1-q'')(-f_1) < U_{1'}$ (Campaign) $= q'(r_1 - c_1 - i) + (1-q')(-f_1 - i)$. Its strong counterpart, by contrast, would strongly prefer to do so: U_1 (No campaign) $= (1-q'')(-f_1) < U_1$ (Campaign) $= q'(r_1 - i) + (1-q')(f_1 - i)$. This also implies that $1 - p' = 0$ and $p' = 1$. As a consequence of this belief, player 2 would strongly prefer to accept after receiving a message. If $q' = 1$ accordingly, player 1 starts a campaign with probability $s = 1$ contradicting the initial assumption $s = 0$.

Case 2: $0 < s < 1, t < 1$

Case 2a ($0 < s < 1$, $t = 0$) can be discarded by considering that $p' = 1$ in consequence. This would again lead to $q' = 1$ and $s = 1$. Case 2b ($0 < s < 1$, $0 < t < 1$), by contrast, assumes that both types of governments mix their strategies. This case can be excluded because such behaviour would lead to different reactions towards the two governments by player 2. More precisely, $U_{1'}$ (Campaign) = q' $(r_1 - c_1 - i) + (1 - q')(f_1 - i)$ and $U_{1'}$ (No Campaign) = $(1 - q'')$ f_1 leads to $q' = (i - q''f_1)/(r_1 - c_1 - f_1)$ while U_1 (Campaign) = q' $(r_1 - i) + (1 - q')(f_1 - i)$ and U_1 (No Campaign) = $(1 - q'')f_1$ invokes $q' = (i - q''f_1)/(r_1 - f_1)$. This is not possible at the same information set.

Case 3: $s = 1, t < 1$

I first analyse the possibility that 1' employs a mixed strategy (Case 3a). This behaviour is a consequence of 2 randomizing between accepting and rejecting. Because $U_{1'}$ (Campaign) = q' $(r_1 - c_1 - i) + (1 - q')(-f_1)^{-i} = U_{1'}$ (No Campaign) $= -f_1$ by consequence, I can derive that $q' = i/(r_1 - c_1 + f_1)$. By using Bayes' rule, I can furthermore calculate t. The equation U_2 (accept) $= p'b_2 + (1 - p')(-c_2)$ $= U_2$ (reject) $= p'$ $(-r_2) + (1 - p')(-r_2)$ first circumscribes $p' = [c_2 - r_2]/[c_2 + b_2]$ $= p^*$. As p' is also equal to $p/(p - (1 - p)t)$, I establish $t = [p/(1 - p)]$ $[(r_2 + b_2)/(r_2 - c_2)]$. Given the strategies and the supporting belief, p'' has to be zero in return, meaning that the second player always rejects the treaty if it could not observe an information campaign. This completes the description of the first equilibrium which is semi-pooling.

It remains to be shown whether the contingency $s = 1$ and $t = 0$ is a Perfect Bayesian equilibrium. This would be a separating equilibrium, with the priors p' and p'' amounting to 1 and 0 respectively. However, as 2 can never be certain about the identity of the government, such a separation is only possible under complete information. The weak type has therefore an incentive to bluff.

Case 4: $s = t = 1$

If both types of government engage in a campaign with probability 1, there is no possibility that player 2 can distinguish different types. Because no updating according to Bayes' rules takes place, the posterior and the prior after receiving a message are identical ($p' = p$). Player 2 always accepts the offer if $p' = p > p^*$. The second information set is never reached. The belief p'' is thus not defined. In contrast to case 1, there is, however, no incentive to switch to the strategy of No Campaign. This second equilibrium is pooling.

Another pooling equilibrium can be reached under the knife-edge condition that $p = p^*$, fixing $0 < q' < 1$. In this case, it is still a (weakly) dominating strategy for 1' to start a campaign. Having exhausted all possible cases, I have thus derived that only three contingencies are $D - 1$ equilibria. This result is summarized in the following proposition:

> *Proposition 5*: Under incomplete information, the Ratification Game falls into three cases:

EQUILIBRIUM 1: The first equilibrium is semi-pooling and defined by the following strategies and beliefs $s = 1$, $t = [p/(1-p)][(r_2 + b_2)/(r_2 - c_2)]$; $q' = i/(r_1 - c_1 + f_1)$, $q'' = 0$; $p' = p < (c_2 - r_2)/(c_2 + b_2) = p^*$ $p'' = 0$.
EQUILIBRIUM 2: If $p > (c_2 - r_2)/(c_2 + b_2)$, we obtain a pooling equilibrium which is described by the following parameters: $s = 1$, $t = 1$, $q' = 1$, $p' = p > (c_2 - r_2)/(b_2 + c_2) = p^*$.

Under the knife-edge condition $p' = p = (c_2 - r_2)/(b_2 + c_2) = p^*$, another pooling equilibrium exists which I call EQUILIBRIUM 2A: $s = 1$, $t = 1$, $q' = i/(r_1 - c_1 + f_1)$, $p' = p = (c_2 - r_2)/(b_2 + c_2) = p^*$.

Comparative statistics show which parameters affect the likelihood of specific outcomes. It should be noted that this last analytical step only refers to Equilibrium 1. $Pr(r)$ represents the probability that the pivotal constituent mistakenly rejects the Less Treaty, and $Pr(a')$ stands for the likelihood that the mimicking behaviour by 1' is successful:

$$Pr(r) = ps(1 - q') = p\,(1 - [i/(r_1 - c_1 + f_1)])$$

$$Pr(a') = (1-p)tq' = (1-p)[p/(1-p)][(r_2 + b_2)/(r_2 - c_2)][i/(r_1 - c_1 + f_1)]$$

$$= p[(r_2 + b_2)/(r_2 - c_2)][i/(r_1 - c_1 + f_1)].$$

The results in Table A6.1 are derived by partially differentiating the outcome probabilities with respect to different parameters. The table accordingly displays how key variables affect the likelihood of outcomes in Equilibrium 1. A plus sign indicates that the effect is positive and a minus sign that it is negative. A zero stands for the absence of an effect.

Table A6.1 The effect of key parameters on the likelihood of outcomes of the Ratification Game

Outcome	r_1	i	f_1	c_1	b_2	r_2	c_2
$Pr\,(r)$	+	–	+	–	0	0	0
$Pr\,(a')$	–	+	–	–	+	–	–

NOTES

1. While either relying on the Waltzian 'second image' or on the 'second image reversed' (Gourevitch, 1978), the antagonists thereby only repeated an old debate. In the nineteenth century, the German historians von Ranke and Dilthey disagreed about the primacy of either international or domestic politics.
2. This categorization is close to the four dimensions of Randolph (1966), who differentiates between prenegotiation, negotiation, agreement and implementation phases.

3. This game is an extension of Achen (1988) who analysed the interaction between government leaders and influencers as a Stackelberg game. The extension refers to the international dimension.
4. This game resembles, of course, the setter models which study the relationship between an agenda-setter and the voters. The concept has been introduced by Romer and Rosenthal (1978, 1979). Recent contributions study the effects of asymmetric information on the outcome of referendums (Banks, 1990; Lupia 1992).
5. The present analysis has had to rely on informal accounts because the official negotiation positions have not yet become public. I exclude Liechtenstein from the analysis because it only became a full EFTA member in 1991.
6. The *acquis communautaire* embodies Community rules in the four relevant areas (free circulation of goods, persons, capital and services) laid down by the EC Treaties, by Council and Community legislation and by case law.
7. This list is adapted from an unpublished document obtained during interviews at the EFTA, Geneva, May 1991.
8. A popular initiative forced Liechtenstein to introduce a new article into its constitution, aiming specifically at the ratification of the EEA agreement and stipulating an optional referendum for all international treaties.
9. In contrast to the game described in Schneider and Cederman (1992), no reference to an equilibrium refinement such as universal divinity is necessary. A further difference is that signalling is always costly in the present model. For a technical introduction to other signalling games in political science, see Banks (1992). Based on the pioneering work by Spence (1973), the theoretical foundations for this special class of limited-information models are largely due to Kreps and his collaborators (Kreps and Wilson, 1982, and especially, Cho and Kreps, 1987).

REFERENCES

Achen, Christopher (1988), 'A State with Bureaucratic Politics is Representable as a Unitary Rational Actor', presented at the annual meeting of the American Political Science Association.

Allan, Pierre (1984), 'Comment négocier en situation de faiblesse? Une typologie des stratégies', *Schweizerisches Jahrbuch für Politische Wissenschaft*, **24**, 223–37.

Allison, Graham (1971), *Essence of Decision: Explaining the Cuban Missile Crisis*, Boston, Mass.: Little, Brown.

Almond, Gabriel A. (1989), Review article: 'The International–National Connection', *British Journal of Political Science*, **19**, 237–59.

Aumann, Robert J. (1974), 'Subjectivity and Correlation in Randomized Experiments', *Journal of Mathematical Economics*, **1**, 67–96.

Banks, Jeffrey S. (1990), 'Monopoly Agenda Control and Asymmetric Information', *Quarterly Journal of Economics*, **105**, 445–64.

—— (1992), *Signaling Games in Political Science*, Chur: Harwood Academic Publishers.

—— and Sobel, Joel (1987), "Equilibrium Selection in Signaling Games', *Econometrica*, **55**, 647–62.

Baron, David P. and Ferejohn, John A. (1989), 'Bargaining in Legislatures', *American Political Science Review*, **83**, 1181–1206.

Bendor, Jonathan and Hammond, Thomas H. (1992), 'Rethinking Allison's Model', *American Political Science Review*, **86**, 301–22.

Brams, Steven J. (1990), 'Games Nations Play in Arms-control Negotiations', New York University, typescript.

Bueno de Mesquita, Bruce and Lalman, David (1992), *War and Reason: A Confrontation between Domestic and International Imperatives*, New Haven, Conn.: Yale University Press.

Cho, In-Koo and Kreps, David M. (1987), 'Signaling Games and Stable Equilibria', *Quarterly Journal of Economics*, **102**, 179–221.

Delors, Jacques (1989), 'Statement on the Broad Lines of Commission Policy', *Bulletin of the European Communities*, Supplement 1/89.

—— (1990), 'Address by President Delors to the European Parliament Presenting the Commission's Programme for 1990', *Bulletin of the European Communities*, Supplement 1/90.

Deutsch, Karl W. *et al.* (1957), *Political Community and the North Atlantic Area: International Organization in the Light of Historical Experience*, Princeton, N.J.: Princeton University Press.

Dupont, Cédric (1992), 'Succès avec la SDN, échec avec l'EEE? Résistances internes et négotiation internationale', *Schweizerisches Jahrbuch für Politische Wissenschaft*, **32**, 249–72.

George, Stephen (1985), *Politics and Policy in the European Community*, London: Oxford University Press.

Gourevitch, Peter (1978), 'The Second Image Reversed: the International Sources of Domestic Politics', *International Organization*, **32**, 881–912.

Groom A. J. R. and Heraclides, Alexis (1985), 'Integration and Disintegration', in Margot Light and A.J.R. Groom (eds), *International Relations: A Handbook of Current Theory*, London: Frances Pinter.

Haas, Ernst B. (1958), *The Uniting of Europe: Political, Social and Economical Forces, 1950–1957*, London: Stevens.

—— (1964), *Beyond the Nation-state: Functionalism and International Organization*, Stanford, Cal.: Stanford University Press.

Iida, Keisuke (1991), 'Two-level Games with Uncertainty: an Extension of Putnam's Theory', Princeton University, typescript.

Janssen, Joseph I.H. (1991), 'Postmaterialism, Cognitive Mobilization and Public Support for European Integration', *British Journal of Political Science*, **21**, 433–68.

Kratochwil, Friedrich and Ruggie, John Gerard (1986), 'International Organization: a State of the Art or an Art of the State', *International Organization*, **40**, 753–75.

Kreps, David M. (1989), 'Out-of-equilibrium Beliefs and Out-of-equilibrium Behavior', in Frank Hahn (ed.), *The Economics of Missing Markets, Information, and Games*, Oxford: Clarendon Press.

—— and Wilson, Robert (1982), 'Sequential Equilibria', *Econometrica*, **50**, 863–94.

Laver, Michael and Shepsle, Kenneth A. (1990), 'Government Coalitions and Intraparty Politics', *British Journal of Political Science*, **25**, 489–507.

Lax, David A. and Sebenius James K. (1991), 'Negotiating through an Agent', *Journal of Conflict Resolution*, **35**, 474–93.

Luif, Paul (1990), 'Austria's Application for EC Membership: Historical Background, Reasons and Possible Results', in Finn Laursen (ed.), *EFTA and the EC: Implications of 1992*, Maastricht: European Institute of Public Administration.

Lupia, Arthur (1992), 'Busy Voters, Agenda Control, and the Power of Information', *American Political Science Review*, **86**, 390–403.

McGinnis, Michael and Williams, John T. (1991), 'Configurations of Cooperation: Correlated Equilibria in Coordination and Iterated Prisoner's Dilemma Games', paper presented at the 25th annual meeting of the Peace Science Society (International).

Moravcsik, Andrew (1991), 'Negotiating the Single European Act: National Interests and Conventional Statecraft in the European Community', *International Organization*, **45**, 19–56.

Morrow, James D. (1991), 'Electoral and Congressional Incentives and Arms Control', *Journal of Conflict Resolution*, **35**, 245–65.

Mutimer, David (1989), '1992 and the Political Integration of Europe: Neo-functionalism Reconsidered', *Revue d'intégration européenne*, **13**, 75–101.

Putnam, Robert D. (1988), 'Diplomacy and Domestic Politics: the Logic of Two-level Games', *International Organization*, **42**, 427–60.

Randolph, Lillian (1966), 'A Suggested Model of International Negotiation', *Journal of Conflict Resolution*, **10**, 344–53.

Richards, Diana, Morgan, T. Clifton, Wilson, Rick K., Schwebach, Valerie L. and Young, Garry D. (1992), 'Good Times, Bad Times and the Diversionary Use of Force: a Tale of Some Not-so-free Agents', paper presented at the annual meeting of the International Studies Association.

Rochester, J. Martin (1986), 'The Rise and Fall of International Organization as a Field of Study', *International Organization*, **40**, 849–94.

Romer, Thomas and Rosenthal Howard (1978), 'Political Resource Allocation, Controlled Agendas, and the Status Quo', *Public Choice*, **33**, 27–44.

—— (1979), 'Bureaucrats versus Voters: On the Political Economy of Resource Allocation by Direct Democracy', *Quarterly Journal of Economics*, **93**, 563–87.

Rubinstein, Ariel (1982), 'Perfect Equilibrium in a Bargaining Model', *Econometrica*, **50**, 79–100.

Scharpf, Fritz W. (1988), 'The Joint-Decision Trap: Lessons from German Federalism and European Integration', *Public Administration*, **66**, 239–78.

Schelling, Thomas (1960), *The Strategy of Conflict*, Cambridge, Mass.: Harvard University Press.

Schneider, Gerald (1992), 'Bremsen und Beschleunigen in Integrationsprozessen. Eine spieltheoretische Analyse zwischenstaatlicher Annäherungsstrategien am Beispiel der EWR-Verhandlungen', *Schweizerisches Jahrbuch für Politische Wissenschaft*, **32**, 273–96.

—— and Cederman, Lars-Erik (1992), 'The Change of Tide in Political Cooperation: a Limited Information Model of European Integration', University of Michigan, typescript.

Select Committee on the European Communities (1990), *Relations Between the Community and the EFTA*, House of Lords, Session 1989–90, 14th Report, London: Her Majesty's Stationery Office.

Spence, Michael (1973), 'Job Market Signaling', *Quarterly Journal of Economics*, **87**, 355–74.

Tsebelis, George (1990), *Nested Games: Rational Choice in Comparative Politics*, Berkeley, Cal.: University of California Press.

Wallace, William (1990), 'Introduction: The Dynamics of European Integration', in William Wallace (ed.), *The Dynamics of European Integration*, London: Frances Pinter.

Waltz, Kenneth W. (1979), *Theory of International Politics*, Reading, Mass.: Addison-Wesley.

7. Domestic politics and international negotiations: a sequential bargaining model

Cédric Dupont*

1 DOMESTIC POLITICS AND INTERNATIONAL NEGOTIATIONS

It is now largely accepted that domestic constraints are necessary to understand state behaviour at the international level. This is particularly true for the study of international negotiations.[1] Classical work illuminated the various aspects of these factors.[2] Domestic constraints may be a bargaining asset and may enable a government to stick firmly to an otherwise untenable position (Schelling, 1960). They may also be a handicap, because any negotiating position will be under severe attack and internal divisions may always be exploited by others (Iklé, 1961).

Recent work has focused on the observation first stated by Walton and McKersie (1965) in the area of labour negotiations: that negotiators are the recipients of two sets of demands – one across the table and one from their own constituencies or audiences.[3] Putnam (1988) popularized this work with his 'two-levels games' metaphor. Negotiators consciously take part in a domestic game and an international game simultaneously, and do act to optimize their positions both at home and at the international level. This conceptualisation of the interplay of domestic politics raises two issues. The first is a methodological one: as Scharpf (1991, p. 293) nicely shows, the introduction of some uncertainty into this framework quickly leads to an indeterminacy trap. It becomes impossible to derive statements on the likely outcomes of the whole process. The second is substantive: do actors really consciously act as if they are part of two simultaneous games? Empirical work has been inconclusive up to now.

* I should like to thank Pierre Allan, Brook Boyer, Max Cameron, Antoine Fleury, Simon Hug, Christian Schmidt and the participants in the Political Science Colloquium at the Graduate Institute of International Studies, Geneva, for their helpful comments on various drafts of this article.

Figure 7.1 Analytical framework

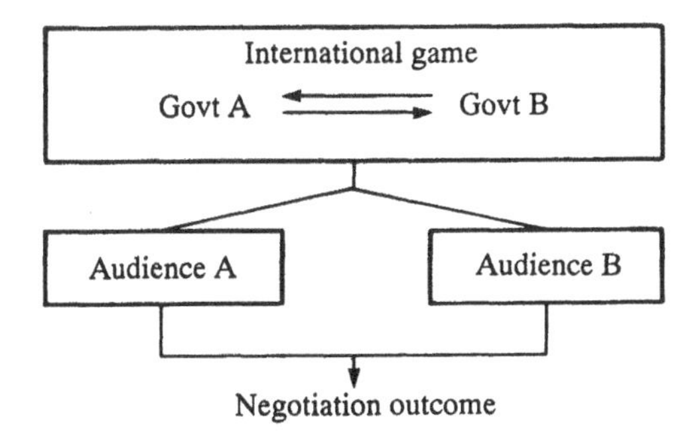

Previous work on non-crisis negotiations (Aggarwal, 1989; Aggarwal and Allan, 1992 and ch. 2 in this volume; Dupont, 1991, 1992; Iida, 1991, Schneider, ch. 6 in this volume) has tried to address these problems either through integrated frameworks with stylization of domestic or international processes, or through segregated games. In this chapter I follow the first option (see Figure 7.1). Strictly speaking, there is only one game, the international game. Domestic politics consists in a ratification process once the international game has resulted in an agreement. This process can be a popular referendum, a parliamentary vote or an approval by small committees. I envisage it within the median voter framework (Black, 1958) and exclude any strategic voting behaviour.

Players in the international game incorporate into their strategic computations constraints coming from the other actor and from domestic audiences. Players are not myopic but anticipate in the international game subsequent domestic ratification processes. In order to do so, we assume that they have a good knowledge of their own domestic constraints, or median voter, and at least some beliefs on the other's. The process and outcome of negotiations are thus determined by the interplay of domestic politics, international structures and mutual perceptions. My main hypothesis is that negotiators consciously exploit domestic ratification constraints to secure international concessions, and they do so strategically in conjunction with international structural determinants and individual cognitive factors.

In this chapter I adopt what Myerson (1992) calls a modelling dialogue. I consider a simple bargaining model, carefully specify it and analyse it under various information conditions. This helps to illuminate the potential impact of the main determinants, namely the configuration of power at the international level, the domestic ratification constraints and the perceptions held by the actors. Successive analyses of negotiations between Switzerland and the League of Nations about Swiss adhesion to the League, and of negotiations between the European Communities and the European Free Trade Association (EFTA) about the creation of a unified market establish a direct correspondence between the model, the propositions I derive from it, and empirical evidence.

2 A GAME THEORETIC SEQUENTIAL BARGAINING MODEL

Game theory is best suited for illuminating the problem of strategic interaction inherent in the analytical framework presented above.[4] Among the vast and diverse game-theoretic literature on bargaining, I adopt the alternating-offers structure popularized by Rubinstein (1982), and develop a two-period model with incomplete information.[5] With this model we can address such basic questions as: (a) what is the outcome of the process?; (b) what are the factors which influence this outcome?; (c) is the outcome stable or unstable?; (d) what conditions will lead to an agreement or to a breakdown?; and (e) what are the effects of varying the distribution of information among players? The ensuing discussion of these questions begins with a description of the structure of the game.

2.1 The Structure of Bargaining

The situation we model is the following.[6] Two players, a 'buyer' A and a 'seller' B, bargain over the price of a given 'good'.[7] An agreement is a price $p \in P$ at a time $t \in T$. The players' preferences are opposed. The seller prefers a higher price whereas the buyer prefers a lower price. Each player is concerned only about what he receives or about what he pays. That is, player A prefers $p_i \in P$ to $p_j \in P$ if and only if $p_i < p_j$, whereas player B prefers $p_i \in P$ to $p_j \in P$ if and only if $p_i > p_j$. The zone of agreement (Raiffa, 1982, pp. 46–7) P is bounded by the actors' 'reservation prices' or valuations, a for actor A and b for actor B (see Figure 7.2). These valuations define the types of actors under conditions of incomplete information. This implies that if $a < b$, the set of agreement P is an empty set.[8]

Figure 7.2 Zone of agreement

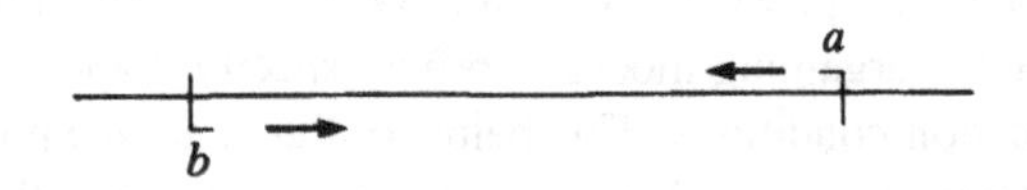

The bargaining procedure is as follows. The players can make an offer only at times in the finite set $T = \{1,2\}$. In period 1, actor A proposes an agreement

Figure 7.3 A simple bargaining game with alternating offers

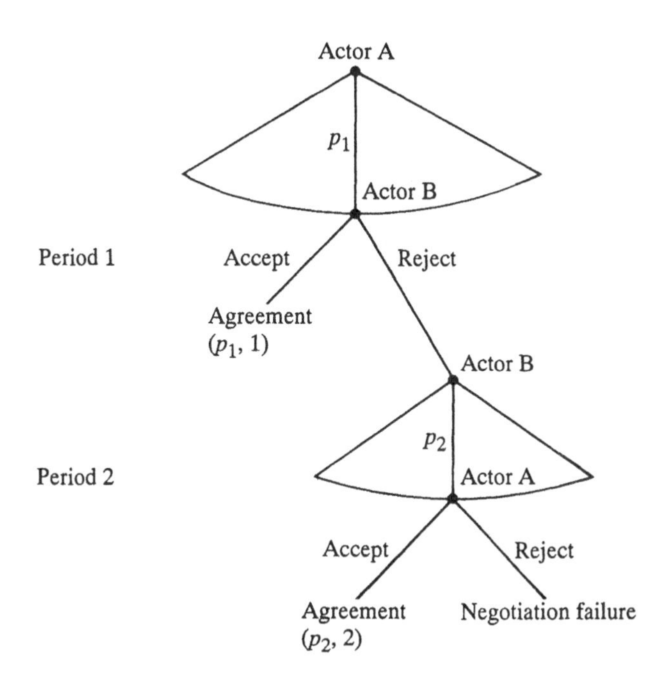

and actor B either accepts the offer or rejects it. If the offer is accepted, the negotiation process ends, and the actors then turn back to their domestic constituent to have the agreement ratified. If the offer is rejected, the game moves to period 2; in this period actor B offers an agreement, which actor A may accept or reject. If she accepts it, then the negotiation ends on the agreement proposed by actor B. If she rejects it, then the negotiation ends on a disagreement and both actors receive their disagreement payoff. At all times, each player knows all his previous moves and all those of the other player. We model this negotiation process as an extensive game pictured on Figure 7.3. We have attached to the terminal nodes where an agreement is reached a pair of the form (p, t), giving the agreement price and the time period at which it is reached. Note that this game tree is infinite in respect to the choices of the players. At any node where an actor makes an offer, there is a continuum (cone-shaped in Figure 7.3), rather than a finite number of choices. Each possible proposal by one player leads to a decision node for the other player, at which she (he) accepts or rejects the proposal. Therefore, it is impossible to represent all the alternatives and the subsequent options of the other player, and we have chosen one 'hypothetical' choice for player A in period 1 and one for player B in period 2. Note also that outcomes are continuously variable: actors do not choose between distinct and distributely different solutions. Lumpy outcomes might seem more realistic but in fact outcomes can be made continuously variable through side payments or package deals.

This is a simple setting, but it is rich enough to highlight some interesting theoretical perspectives. It is a non-cooperative bargaining model that captures both the fact that negotiation involves a succession of steps and that actors typically are uncertain about the value to others of reaching an agreement. And domestic political variables can be incorporated into this framework.

More precisely, I choose to link domestic politics with the minimum requirement levels, or reservation prices. This reflects the position of the pivotal constituent (Black, 1958) in the constitutional vote on the issue being negotiated.[9] For actor A, the reservation value is thus the highest price that will be accepted in the popular vote on the issue at hand. Any more costly agreement will be rejected. For actor B, it is the lowest price that would be accepted in a popular vote. If we restrict the possible agreement to $\mathfrak{R}^+$, the 'win-sets' of player A and player B are thus $[0, a]$ and $[b, \infty)$, respectively.

In the substantive context of international negotiations, this theoretical framework helps to illuminate a large set of non-crisis interactions. Actors A and B can be seen as two countries bargaining about regulation of transnational pollution, immigration policy or extension of land lease for military bases. For the case of transnational pollution, for instance, suppose actor A is a country that wants its neighbouring state, actor B, to reduce the activities of polluting firms along their common border. In order to induce B, actor A will have to pay a price, for instance give wage compensation to the workers of the polluting firms. Domestic constituents in country A set the upper limit of this price. On the other hand, in country B domestic constituents will influence the minimum acceptable compensation offer from country A. Negotiations will start only if both countries' reservation levels are compatible. As my empirical illustrations in section 4 will show, the model formalizes other negotiations where actors need not be single countries but can also be international organizations bargaining over issues such as membership application or economic integration. To go further and analyse the players' choices, we next specify their preferences over the possible outcomes of the negotiation process, and their information on these preferences.

2.2 Preferences

P1 (Price preference)

We assume that each actor i = A, B has a continuous utility function u_i: $[b,a] \to \mathfrak{R}$ such that for any $p \in [b,a]$ we have:

$$u_A(p) = a - p$$

$$u_B(p) = p - b$$

The assumption of continuity might appear as a serious limitation to the application of economic bargaining models to political situations. But in most international negotiations, various piecemeal concessions can be arranged such that utility levels for the negotiating parties vary along a continuum.

P2 (Disagreement payoffs)
In case of disagreement, D, both actors get a zero payoff. The disagreement is thus the worst outcome.

This implies that no actor has an incentive to deliberately strive for a disagreement, and that no player has true alternative or valuable outside options.[10]

P3 (Time preference)
For any $t \in T$, $s \in T$, and $p \in [b,a]$, we have $u_i(p,t) \geq u_i(p,s)$ if $t < s$ with strict inequality if $u_i > 0$.

We can now define a continuous utility function U_i: $[b,a] \times T \cup \{D\} \rightarrow \mathfrak{R}$ such that for $i = A, B$, $t \in \{1,2\}$, and $p \in [b,a]$, we have: $U_i(p,t) = \delta_i^{t-1} u_i(p)$ where δ_i is the discount factor of each actor.

Information on preferences
Time preferences and disagreement payoffs are common knowledge among the players. For price preferences, information on reservation prices may vary, and we examine the implications of different information distributions in section 3.

2.3 Strategies and Equilibrium Definition

A strategy of a player specifies a behaviour at every node of the game at which she (he) has to make a decision. In period 1, player A's strategy is an offer $p_1(a) \in [b,a]$ and player B strategy is a response $r_1(p_1,b) \in \{0,1\}$.[11] In period 2, player B's strategy is an offer $p_2(b) \in [b,a]$ and player A's strategy is a response $r_2(p_2,a) \in \{0,1\}$.

An equilibrium is a set of strategies (p_1,p_2,r_1,r_2) forming a subgame perfect Nash equilibrium for the complete information situation. For cases of incomplete information, this set of strategies and the system of beliefs attached to the various types of actors form a perfect Bayesian equilibrium (Fudenberg and Tirole, 1991, pp. 321ff). No player can gain by deviating to another strategy, given his beliefs and the strategies of the other player, the updating of beliefs being performed according to Bayes' rule wherever possible. In this simple setting, player A's second period equilibrium strategy is easily computed: she accepts an offer p_2 if and only if p_2 does not exceed a. More formally: $r_2(p_2,a) = 1$ if $p_2 \leq a$ and $r_2(p_2,a) = 0$ if $p_2 > a$.[12]

3 ANALYSIS OF THE BARGAINING MODEL UNDER VARIOUS INFORMATION DISTRIBUTIONS

We begin with a situation where both actors' valuations are known to the other one. We then present an asymmetric case where actor B does not know actor's A valuation, and finally move on to a case where both actors know their valuations but not their opponent's.

3.1 Complete Information about the Actors' Valuations

We first state the following proposition:

> *Proposition 1*: Under complete information, the bargaining game has the following subgame perfect equilibrium: in period 1, actor A makes the offer $v = b(1 - \delta_B) + \delta_B a$ and B accepts any offer $p_1 \geq v$. In period 2, B makes the offer $p_2 = a$ and A accepts any offer $p_2 \leq a$.
> PROOF: We solve the game using backward induction. We begin in period 2: actor A accepts any price $p_2 \leq a$. Hence B offers a which gives the following payoffs to the two players: $U_B(a,2) = \delta_B(a - b)$ and $U_A(a,2) = 0$
> Going backward, in period 1, player B accepts any price p_1 such that $U_B(p_1,1) \geq U_B(a,2)$. Thus player A offers p_1 such that $p_1 - b = \delta_B(a - b)$ which is equivalent to $p_1 = b(1 - \delta_B) + \delta_B a = v$. QED

If we assume complete information about the players' valuation, the bargaining mechanism is fully efficient: an agreement is reached in period 1 and there is no loss in the gains of trade. Due to the structure of the bargaining, the driving force is the time preference of actor B. If the latter is patient, $\delta_B = 1$, he is able to get the price a which leaves actor A with a payoff of zero; if B is fully impatient, $\delta_B = 0$, then A offers b in period 1 and B accepts it.

Our very simple bargaining model clearly shows a last-mover advantage if actor B is patient and a first-mover advantage if actor B is rather impatient. This highlights the crucial importance of the structure of bargaining and the difficulty to come up with a universal setting. We present the complete information outcome as a benchmark to evaluate the results of the incomplete information cases.

What can we say if domestic political constraints restrict the feasible set of agreements, or in other words, decrease a or increase b?

- If we make actor A's reservation price after a severing of domestic constraints equal to some $a' < a$ while keeping constant B's reservation price at b, the outcome is the following: in period 1, actor A makes the offer $p_1 = b(1 - \delta_B) + \delta_B a'$ and B accepts it. In period 2, B makes the offer $p_2 = a'$ and A accepts it. Logically, the agreement price is lower than in the unconstrained case. It follows that actor B's utility is lower for any δ_B. For player A, we cannot compare utilities of two different players. Note, however, that severing domestic constraints does not change the picture if $\delta_B = 1$ but if $\delta_B < 1$ his utility is also negatively affected. This could be surprising but it highly depends on the specific form of the utility function. Internationally, the domestically constrained government cannot exploit all the potential gains from trade but may be better off at the domestic level.
- If we increase b to some b', keeping a constant, the outcome would be: in period 1, actor A makes the offer $p_1 = b'(1 - \delta_B) + \delta_B a$ and B accepts it. In period 2, B makes the offer $p_2 = a$ and A accepts it.

  If player B is totally patient, having more severe domestic constraints does not change the agreement price. If this is not the case, then agreement occurs at a higher price, and actor A is negatively affected by it.
- With both constraints changing, the agreement price is $p_1 = b'(1 - \delta_B) + \delta_B a'$ depending on the value of δ_B this can be higher or lower than the case with the constraints equal to a and b respectively.

3.2 One-sided Incomplete Information: Two Types of Actor A

Let us introduce some uncertainty about one player's reservation price. We suppose that player B is uncertain about the constraints actor A faces domestically. We call type $\underline{a}$ actor A with 'tight' domestic constraints and type $\bar{a}$ actor A with 'loose' domestic constraints. Player B's beliefs are $(x, 1 - x)$ regarding type $\underline{a}$ and type $\bar{a}$ respectively, and this pair is common knowledge. For matters of clarity I present the results for the case where $(x, 1 - x) = (\frac{1}{2}, \frac{1}{2})$.[13] Actor B's valuation is common knowledge.

Before we turn to the description of equilibrium behaviour it is useful to note that we consider two situations: one with a 'soft' actor B and one with a 'hard' country B. What do we mean by that? A soft actor B does not make a second-period offer that could be rejected by one type of actor A, specifically the tightly constrained type $\underline{a}$ whereas the 'hard' actor B prefers to get the most favourable agreement with type $\bar{a}$ even if he knows that the other type will not be able to accept it.[14] Disagreement is in some sense too inefficient for a 'soft', or dovish, actor B. This is not the case for a 'hard', or hawkish, player B. For the specific case in which the beliefs are $(\frac{1}{2}, \frac{1}{2})$ we have a dovish B if $\bar{a} > \underline{a} > (b + \bar{a})/2$ and a hawkish B if $\bar{a} > (\bar{a} + b)/2 > \underline{a}$ (see Figure 7.4 for a graphical illustration of these definitions).

Figure 7.4 Definitions of actor B under incomplete information

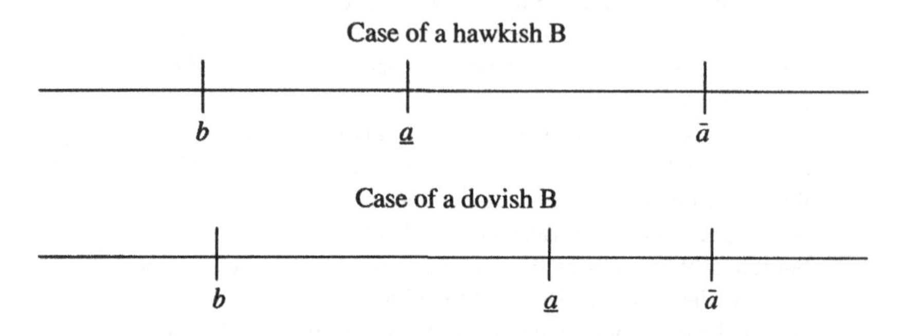

For simplicity I shall ignore borderline cases such as $\underline{a} = (b + \bar{a})/2$; the following statements thus hold for almost all parameter values.

> *Proposition* 2: Under incomplete information about actor A's valuation, with prior $(\frac{1}{2},\frac{1}{2})$, we have the following equilibrium behaviour:
>
> - Case 1: If player B is soft, that is $\bar{a} > \underline{a} > (b + \bar{a})/2$, we have a pooling equilibrium. In period 1 both types of A offer $p_1 = \dot{v} = \delta_B\underline{a} + b(1 - \delta_B)$. Actor B accepts any offer $p_1 \geq \dot{v}$. In period 2, B offers $\underline{a}$; type $\underline{a}$ accepts any offer $p_2 \leq \underline{a}$ and type $\bar{a}$ accepts any offer $p_2 \leq \bar{a}$.
> - Case 2: If player B is hard, that is $\bar{a} > (\bar{a} + b)/2 > \underline{a}$, we have the following: **2a:** if $\delta_B \leq 2(\underline{a} - b)/(\bar{a} - b)$: pooling equilibrium. In period 1, both types of A offer $\tilde{v} = \frac{1}{2}\delta_B\bar{a} + b(1 - \frac{1}{2}\delta_B)$ and B accepts any offer greater or equal to $\tilde{v}$. In period 2, B offers $\bar{a}$, type $\underline{a}$ accepts any offer $p_2 \leq \underline{a}$ and type $\bar{a}$ any offer $p_2 \leq \bar{a}$.
>   **2b:** if $\delta_B > 2(\underline{a} - b)/(\bar{a} - b)$: there is no perfect Bayesian equilibrium in pure strategies.
>
> PROOF: see Appendix.

The introduction of some uncertainty significantly affects the negotiation process. Incomplete information brings inefficiency in the bargaining mechanism. Agreement may not be reached in the first period and may even not occur, especially when player B is hawkish. Severe domestic constraints on player A's side do not necessarily lead to a better agreement because B prefers in some cases to forgo an agreement with type $\underline{a}$ for a more beneficial agreement with type $\bar{a}$ only. This implies on the one hand that a small exaggeration by player A of her reservation price has a high probability of being successful, whereas a large exaggeration might be counterproductive. On the other hand, a truly constrained actor A will be able to win concessions only if she is able to credibly persuade B of her type. Due to the structure of the bargaining this is very difficult. Type $\underline{a}$ has no incentive to signal his type and always prefer to free-ride on type a.[15] Player B has thus no reason to believe that he is really facing the tightly constrained actor A.

3.3 Two-sided Incomplete Information: Two Types of A, Two Types of B

Let us now introduce uncertainty about player B's domestic constraints. We call type $\underline{b}$ actor B with 'loose' domestic constraints and type $\bar{b}$ actor B with 'tight' domestic constraints. Player A's beliefs are $(\frac{1}{2},\frac{1}{2})$ regarding type $\underline{b}$ and type $\bar{b}$, respectively. We keep the two previously defined types of player A and actor B's associated set of beliefs $(\frac{1}{2},\frac{1}{2})$. Both actors' beliefs are common knowledge.

Regarding the 'nature' of actors (dovish or hawkish), each type of each player may be dovish or hawkish. For player B we keep the same definition as in section 3.2. For the other player, a dovish A, type $\underline{a}$ or type $\bar{a}$ prefers to have an agreement at $\bar{b}$ with both types of B rather than an agreement at $\underline{b}$ with only the loosely constrained type $\underline{b}$, whereas a hawkish prefers to have an agreement only with the loosely constrained type of B. For the specific pair of beliefs $(\frac{1}{2},\frac{1}{2})$ we have a hawkish A if $\bar{b} > (\underline{b} + a)/2 > \underline{b}$ and a dovish A if $(\underline{b} + a)/2 > \bar{b} > \underline{b}$ $(a = \underline{a}, \bar{a})$. We consider cases where there is always a potential gain from trade, that is where $\underline{a} > \bar{b}$ and ignore borderlines (see Figure 7.5 for a graphical illustration of a case with one dovish and one hawkish type of each player).

Figure 7.5 A situation with one hawkish and one dovish type of each player

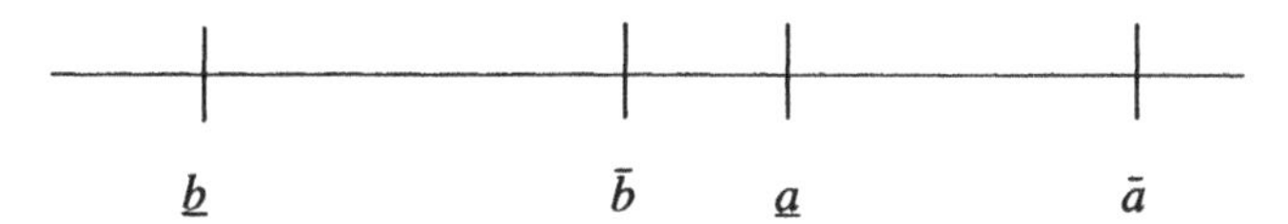

The following statements thus hold for almost all values of the parameters:

Proposition 3: Under two-sided incomplete information, with two types of each actor, and with systems of beliefs $(\frac{1}{2},\frac{1}{2})$ and $(\frac{1}{2},\frac{1}{2})$, we find the following results regarding equilibrium behaviour:

- Case 1: Two types of B are dovish:
  1a: Two types of A are hawkish: – if

$$\delta_A \geq \frac{\left(\bar{a} + \underline{b} - 2\bar{b}\right) + \delta_B\left(2\bar{b} - \underline{b} - \underline{a}\right)}{\bar{a} - \underline{a}},$$

pooling equilibrium at $\check{v} = \underline{b}(1 - \delta_B) + \delta_B \underline{a}$. Type $\bar{b}$ accepts any offer $\geq \hat{v} = \bar{b}(1 - \delta_B) + \delta_B \underline{a}$ and type $\underline{b}$ any offer $\geq \check{v}$. In period 2, both types of B offer $\underline{a}$; otherwise, no equilibrium.

1b: Two types of A are dovish: pooling equilibrium at $\hat{v}$; type $\bar{b}$ accepts an offer $\geq \hat{v}$ and type $\underline{b}$ any offer $\geq \check{v}$. In period 2, both types of B offer $\underline{a}$.

1c: Type $\underline{a}$ hawkish, $\bar{a}$ dovish: – if

$$\delta_A \geq \frac{\left(\bar{a} + \underline{b} - 2\bar{b}\right) + \delta_B\left(2\bar{b} - \underline{b} - \underline{a}\right)}{\bar{a} - \underline{a}},$$

pooling equilibrium at $\check{v}$; otherwise, no equilibrium.

- Case 2: Two types of B are hawkish:

  2a: Both types of A are dovish:
  If $\delta_B \leq 2(\underline{a} - \bar{b})/(\bar{a} - \bar{b})$, we have a pooling equilibrium at $\hat{w} = \bar{b}(1 - \frac{1}{2}\delta_B) + \frac{1}{2}\delta_B \bar{a}$, type $\underline{b}$ accepts any offer higher than $\check{w} = \underline{b}(1 - \frac{1}{2}\delta_B) + \frac{1}{2}\delta_B \bar{a}$ and type $\bar{b}$ any offer higher than $\hat{w}$. In period 2, both types of B offer $\bar{a}$.
  If $\delta_B > 2(\underline{a} - \bar{b})/(\bar{a} - \bar{b})$, there is no perfect Bayesian equilibrium in pure strategies.

  2b: Type $\underline{a}$ hawkish, $\bar{a}$ dovish: there is no equilibrium in pure strategies.

- Case 3: Type $\underline{b}$ dovish, and type $\bar{b}$ hawkish:

  3a: If both types of A are hawkish: we have a pooling equilibrium at $\check{v}$ if

$$\delta_B \leq \frac{2\bar{b} - \underline{b} - \bar{a}}{\bar{b} - \underline{b} + \underline{a} - \bar{a}};$$

  if it is not the case, there is no equilibrium in pure strategies.

  3b: If both types of A are dovish: we have a pooling equilibrium at $\hat{w}$ if we have both

$$\delta_B \leq 2(\underline{a} - \bar{b})/(\bar{a} - \bar{b}) \text{ and}$$

$$\delta_B \geq 2(\bar{b} - \underline{b} - \underline{a})/(\bar{b} - \underline{b} - \bar{a} + \underline{a}) \tag{7.1}$$

  Otherwise, there is no equilibrium.

  3c: If type $\underline{a}$ is hawkish and type $\bar{a}$ dovish: there is no equilibrium in pure strategies.

PROOF: see Appendix.

Compared to the case of one-sided incomplete information, inefficiency is even larger with two-sided incomplete information. This is particularly true when the severely constrained players A and B are hawkish.

Regarding the possible exploitation of domestic constraints to induce concessions, we cannot conclude that more severe constraints are always better. A more severely constrained player B can better limit A's incentive to bluff on her constraints, but this might be at the expense of an agreement in some cases. A more severely constrained actor A might get better terms from a dovish B but might fail to reach an agreement with a hawkish B. The result of the process is thus indeterminate for a non-negligible set of parameters. But, contrary to purely empirical analyses, we are able to identify clearly the conditions that lead to this situation and thus highlight the underlying causes.

We now turn to empirical evidence for the theoretical statements we just derived.

4 EMPIRICAL ILLUSTRATIONS

4.1 Switzerland and the League of Nations

I look at negotiations after the First World War between Switzerland and the victorious powers. The main item on the agenda was the question of Swiss neutrality inside the League of Nations. My analysis deals primarily with this core issue, and to a lesser extent with the related issue of the location of the League's headquarters in Geneva as opposed to Brussels.[16]

I consider the actual bargaining period from February 1919, when the Commission on the League of Nations – the so-called Crillon Commission – met for the first time, to the London declaration of 13 February 1920 when the Council of the League officially accepted a special status for Switzerland inside the League. I focus on the evidence that negotiators consciously use domestic constraints for strategic ends and that they do so in conjunction with international forces and perceptions.

On 16 May 1920 adhesion to the newly born League of Nations was accepted by Swiss voters and cantons in a close vote.[17] This brought to an end a process begun more than one year earlier. To reach an agreement with the Great Powers, Switzerland had to abandon its integral neutrality and replace it with a differentiated one.[18] This still constituted an exceptional situation inside the League. How did the Swiss obtain this privileged arrangement? According to the Federal Council, the explanation is to be found in a combination of domestic specificities, international friendships and negotiators' strategic aptitude.[19] We shall try to shed more light on this interplay with the help of the model developed in section 2. Before deriving predictive statements about the process and contrasting these with the actual bargaining, we first specify the game and identify the various parameters.

Specifying the game and deriving predictions

Who and under what conditions were the actors taking part in the negotiation process? The negotiations over Swiss adhesion to the League were bilateral between Switzerland on the one side, and the victorious belligerent powers grouped in the Supreme Council and later the Council of the League on the other. These two actors were markedly asymmetric. Regarding power distribution, Switzerland was a small non-belligerent country with few resources, highly dependent on the external world as the First World War had just shown. The Swiss faced the coalition of states that had just established themselves as the uncontested dominant powers in the world. As for the institutional nature, we have on the one side a single country, represented by its government and directly influenced by its domestic constituency, and on the other side a coalition of states, influenced both by interstate differences and individual domestic matters.

Such asymmetry is reflected in the respective interest attached to negotiation. Stakes were huge for Switzerland. Politically, the country had to regain some role in world diplomacy and at the same time reaffirm the positive role of its neutrality. Economically, accession to the League of Nations would ensure the opening of large foreign markets to Swiss industries.[20] On the Great Powers' side, evaluating the stakes is more difficult. Obviously, Swiss adhesion to the League was far from being primary, and the interest in negotiations very limited.[21] But this did not preclude any possibility of discussions, given the very specific nature of the issues at hand. Indeed, there existed only one Switzerland and if the Allies wished to see it inside the League they had to take its individual characteristics into account, thus opening a possible range of agreement. This desire existed. President Wilson was probably the foremost advocate of Swiss participation in the League, both on friendly and ethical grounds. Swiss democracy was still considered an ideal, which could contribute to the ethical foundations of the new world society based on Wilsonian principles.[22] This position was shared by British and French leaders.[23] To sum up, the Allies were ready to consider the details of the status of Switzerland inside the League, but there was clearly no possibility of any discussions on the wider aspects of the Peace Conference. Therefore, negotiations only concerned Swiss participation in the League of Nations and not in the Peace Conference.

We now try to establish a close correspondence between the various components of our model and the reality we quickly portrayed. Switzerland is actor A, eager to 'buy' adhesion to the League of Nations from actor B, the Allied Powers. The asymmetry in power is reflected in the structure of the game which favours actor B. This one has a last say in the process since it makes a 'take it or leave it' offer in the second period. The asymmetry in power is also reflected in the value of the discount factor. The Great Powers were not time-impatient since they themselves set the future agenda. Switzerland was much more impatient since it had to respect strictly various deadlines imposed by the Great Powers and by its domestic constituency.

According to the model in section 2, ratification constraints determine the reservation value of each actor. For Switzerland, these consisted of a popular referendum, and accordingly the minimal acceptable conditions were those that would get a majority of the Swiss people to accept the adhesion to the League. These conditions were fairly precisely known (or at least perceived as such) by the Swiss government but relatively unknown to the Allies. For the latter, we do not consider domestic politics of each individual country because the matters under consideration were not significant for the various domestic constituencies. They were sensitive to the big issues, notably the details of war reparations, but surely not to the conditions surrounding Swiss entry into the League. The reservation price is thus operationalized as the median voter in the Supreme Council. We consider that this value was well known both to the Allies and to Switzerland. Indeed, the Swiss had access to leading statesmen and so were very well informed on what was going on between the Allies.[24] At the outset of negotiations in February 1919, the Federal Council could rely on the information collected mainly by its unofficial representative in Paris, William Rappard, but also by the President of the Confederation, Gustave Ador, during his mission in Paris from 21–29 January. We thus have a one-sided incomplete information game, with two types of Switzerland, corresponding to the analysis in section 3.2.

How do we distinguish between the two types of Switzerland? The loosely constrained type, $\bar{a}$ in our theoretical framework, would have been able to adhere to the League without explicit recognition of its neutrality, whereas the tightly constrained type – the 'true' type – was not able to pay such a high price for adhesion. The Swiss needed at least a clear recognition of their neutrality for military measures.

We next specify the nature of actor B, that is, the Supreme Council. Were the Allies hawkish or dovish? This, according to our definition, depended on the beliefs they held over the respective probability of the two types of Switzerland. They seemed to be conscious of Swiss domestic constraints right at the beginning of the negotiation process, but they were far from convinced that these constraints were strictly binding. We can draw no more inferences from empirical evidence, and we therefore consider the even distribution as a good approximation of the deep puzzle regarding Swiss domestic constraints.[25] Given these beliefs, we code the Supreme Council as hawkish. The median position inside the Council was that Switzerland had to pay some price for its accession, namely some restrictions on the scope of neutrality.[26] The Council's reservation price, b, was close to the reservation price of the tightly constrained

Switzerland, type $\underline{a}$. Accordingly, the Supreme Council would be tempted to focus on the loosely constrained type of Switzerland, and forgo any agreement with the other type.

Given these game specifications, we can derive some predictions from the theoretical models depending on prior beliefs. We can see from Proposition 2 that there is no equilibrium in pure strategies with a hawkish Supreme Council. The two types of Switzerland do not pool, and cannot be strictly separated. A whole range of outcomes is thus possible, including delays in agreement or even disagreement. This last result is most probable for the tightly constrained Switzerland, type $\underline{a}$. This type cannot credibly signal its identity to the Supreme Council. No stable result is expected unless there is a game change. This change can only come from a major revision in the Allies' beliefs, and thus we expect Swiss efforts to concentrate on this aspect. If information becomes complete, agreement is no longer problematic but immediate at Switzerland's reservation price.

If actors consciously incorporate internal constraints the way we theoretically assume they do, are these rough predictions consistent with the historical record? We now address this question by presenting the actual bargaining.

The bargaining

From February 1919 to June 1919: domestic politics kept aside From November 1918 to the beginning of February 1919, Switzerland was unsuccessful in becoming a member of the Paris Peace Conference. It had hoped to have a say in the Conference through a carefully prepared draft covenant. The draft was submitted to the powers in February 1919. It was praised as an excellent piece of work but its influence was 'practically nil' (Meier, 1970, p. 110). The result was that Switzerland was left aside, facing the possibility of having to accept without any modification the outcome of the Peace Conference concerning the organization of the League of Nations. This was a serious concern for the Federal Council given the constitutional constraint to get popular approval for accession to the League.[27] The Swiss people were deeply attached to their neutrality and it was doubtful that they would have accepted entry to the League of Nations if it implied giving up this deeply rooted policy.[28]

With no alternative but to negotiate, the Federal Council had to find a way out of this potential deadlock. Domestic constraints were not considered at that time to be a potentially useful strategic tool. Rather, they were considered to be a disadvantage compared to nations without any popular ratification processes.[29] Consequently, Swiss negotiators preferred to use historical specificities and American benevolence in order to induce some concessions on the question of neutrality.[30] The Swiss maintained close contacts with the American Legation in Paris because William Rappard had developed a friendly relationship with both President Wilson and his alter ego, Colonel House. He was also in close contact with David Hunter Miller, the chief judicial counsellor of the American delegation.[31] Rappard not only heard everything of importance that was going on in the Conference, but he also found very powerful negotiating 'agents' to defend some of his interests. These 'agents' were, however, very free-wheeling and independent. As a result, the Swiss failed to get any firm commitment about their neutrality during the month of February 1919. Most of the belligerent powers were in favour of Swiss participation in the League. but none really cared about the modalities of adhesion, even if they gave the impression that neutrality was very dear to Switzerland.[32]

American benevolence proved to be more fruitful on the question of the future headquarters of the League. Wilson and the British had for some time discussed Geneva as the possible seat of the new organization; they opposed Brussels, favoured by the French, both for political reasons and moral principles. It was under strong French influence and would have represented a League of Nations equivalent to a club of victors. Thus the Americans encouraged the Federal Council to make an official offer to the Commission. Rappard was prominent in getting the government to react quickly on this matter because he not only saw the material advantages of Geneva but also the strategic dimension of the League's headquarters. Internationally, the wish to guarantee the territorial inviolability of the organization could justify Swiss neutrality; domestically, it would help to materialize the benefits Switzerland could obtain from adhesion.[33]

The Federal Council was much less inclined to take position on the question of the headquarters because it did not want to make a definitive decision about Swiss participation in the League. Pressed by Rappard, by time and by the threat to see the League go to Brussels, the Swiss government finally decided to take action on this issue. The subcommittee appointed to study the question unanimously recommended Geneva as the future headquarters of the League at the fourteenth meeting of the Commission on 10 April 1919. This recommendation was accepted by twelve out of nineteen members, despite French and Belgian efforts to change the tide.[34] During the same meeting, the Commission definitively rejected the amendment to the sanctions article (Article 16), proposed in February by the Swiss in order to preserve Swiss neutrality. No explicit exception would be included in the Covenant for any kind of sanction.[35] The Federal Council had to rely only on informal assurances that some way would be found to preserve neutrality. But this would not be enough to convince first the Parliament, and then the people, to vote in favour of adhesion to the League.

The Federal Council then began to consider the strategic use of these domestic constitutional constraints to get the necessary concessions.[36] This strategy was less risky now that the headquarters had been allocated to Geneva but overt abuse could have reopened the question. It was thus 'tested' on friendly ears, in particular Lord Robert Cecil and Colonel House. The result was disappointing: they understood Swiss domestic concerns but thought that it would not be impossible for Switzerland to adhere to the League under the current conditions. They did not believe these were incompatible with popular demands, and accordingly did not see any need for explicit exceptions.[37] Despite their negative attitude, they suggested a way out of this difficulty: Swiss neutrality could be preserved by Article 21 (the Monroe Doctrine).[38] Wilson endorsed it but advised Rappard to wait for the Covenant to come into force before making any public interpretation along those lines. He assured him that this would be accepted by the powers because the British and the Americans would support it.[39] No prior official declaration would be accepted because it could have created unfortunate misunderstandings.

Negotiations ended on these vague terms. Both sides had come to an agreement but the most sensitive points were still very speculative. Switzerland had apparently received concessions that would ensure popular approval but had no official recognition of it. The position of the Federal Council was indeed very fragile when it turned to the ratification processes.

From July 1919 to February 1920: domestic difficulties and international arrangements Debate in Switzerland had been very limited during the Paris negotiations because the Federal Council tried to discourage it. This fitted the perception that domestic difficulties were a drawback. But it was now imperative to turn to the domestic constituency, first Parliament and then the Swiss people, because no international agreement could be ratified without approval from these two constituents.

In view of the forthcoming discussion in the Federal Assembly, on 4 August 1919 the Federal Council issued a long message in which it carefully explained the international and domestic implications of adhesion to the League.[40] The government officially declared that Switzerland would be able to join the League and keep its neutrality, a fact that could not be found in any official international document. The Federal Council anticipated objection from opponents and clearly committed itself at both the domestic and the international level. Switzerland could not have adhered to the League without at least a differentiated neutrality because the Swiss would not have accepted changing a deeply rooted and firmly established policy for something still vague and uncertain. This was clearly understood abroad, and the Swiss interpretation of the Covenant raised no objections; but renewed efforts by Belgium to get the headquarters of the League prevented the Federal Council from fully exploiting its domestic difficulties.[41] The domestic constraints were set out as a strategic tool but their use was carefully withheld.

Inside Switzerland tensions did not disappear after this message and debates in Parliament, initially scheduled for September, were postponed until November. With the American delegates fully devoted to domestic affairs, the Federal Council gave all its attention in the new Secretariat of the League and the still existing

Supreme Council, and kept them informed upon the attitude of Parliament and public opinion in Switzerland through a daily press report distributed by the Swiss attaché in London, William Martin.[42] This constant inflow of information helped to preserve the goodwill of the powers.

Debates in Parliament revealed a sharp polarization in attitudes towards the League, but the majority of deputies eventually voted in favour of adhesion on 21 November 1919. However the Federal Council was still in an uncomfortable situation. Parliamentary acceptance was dependent upon the explicit recognition by the five Great Powers of Swiss neutrality and upon prior ratification by them.[43] Moreover, Swiss adhesion would probably not respect the time-limit prescribed for original membership and this could reinforce the interests of Brussels as the headquarters of the League. Consequently, the Federal Council faced two alternatives: (a) wait for the entry into force of the Treaty and then make a declaration of adhesion with an explicit mention of Swiss perpetual neutrality; (b) immediately ask for an explicit recognition by the Great Powers of the particular situation of Switzerland. Neither of these two options was safe: the first could have fuelled domestic opposition, and the second could have hurt external susceptibility.

Encouraged by friendly reassurances from Britain, the Federal Council chose to go down the second road and prepared an *aide-mémoire* to be sent to the powers, the Secretariat of the League and the original members.[44] The document aimed to explain the constitutional sources of the probable disregard of the time-limit for original members but did not mention the question of neutrality.[45] The government did not want to reopen a question which had been informally settled.

Contrary to expectations derived from informal contacts between Rappard and various delegations in London, the *aide-mémoire* did not receive the desired reception. The French government had advised the Supreme Council of this matter and the *aide-mémoire* as well as the annexed text of the proceeding of the Swiss Federal Assembly were examined by the Council's legal advisers. Their conclusion was sent to the Swiss government in a Note dated 2 January 1920.[46] On the issue of the time-limit, the Council considered that Swiss constitutional specificities concerned Switzerland only; there would be no exception based on those domestic measures. On the question of neutrality, which was not mentioned in the *aide-mémoire* but in the text of the proceedings of the Assembly, the Council declared that it 'would hold this question for subsequent examination'. The Federal Council's intention was to get an official clarification of previously obtained concessions, but it was not perceived as such by the powers. Indeed, the Council perceived the Swiss *aide-mémoire* as a demand for additional concessions and not as an official clarification note.

Domestically, the Note from the powers was likened to a 'death penalty' (Rappard, 1924, p. 44) for the supporters of adhesion to the League. To get out of this stalemate, the Federal Council reacted quickly and decided to use explicitly the ratification constraints in order to secure definitely and officially the informal assurances received in 1919. The success of this tactic depended on how the Great Powers perceived internal difficulties. The Federal Council proceeded to explain its thorny situation. It first addressed to all the members of the Peace Treaties a memorandum in which the need for clarity as the primary condition for a popular success on the referendum was identified, especially on the question of neutrality.[47] Secondly, an extraordinary mission was sent to Paris to explain the Swiss position to the Supreme Council.

The mission, composed of former Federal Councillor and Swiss President Gustave Ador and Special Judicial Counsellor of the Federal Council Max Huber, was heard on 20 January 1920. Although Ador requested recognition of the exclusion of the right of passage of troops across Swiss territory, he declared that his country was prepared to take part in collective economic sanctions.[48] Subsequent reactions on that and the next day were mainly favourable to special arrangements for Switzerland. The French, who until then had been rather wary, recognised the exceptional position of Switzerland, as did the Italians. On the British side, Lord Curzon was bothered by the time-limit question in that it would create an unfortunate precedent, but he did not object to neutrality on military matters.[49] But these encouraging reactions did not bring any definitive results because the questions were considered to be within the competence of the newly formed Council of the League of Nations.

The mission in Paris was a general repetition of the mission in London as interim host of the Council of the League. Once again Ador asked for the strict necessary arrangements.[50] He grounded these demands on the popular attachment to neutrality. This attachment would determine the issue of the referendum and therefore Switzerland's participation in the League. He thus solicited the help of the delegates if their desire was to see Switzerland inside the League. The next day, 13 February, the Council unanimously passed a resolution specifically protecting Swiss perpetual neutrality and excluding Switzerland from any military action or obligation regarding the passage of foreign troops across its territory (Miller, 1928, pp. 437–8). In addition, the Council accepted that the confirmation of adhesion by the Swiss people should occur beyond the time-limit for original membership.

As the popular referendum would show three months later, these conditions were close to the median voter position. Once the Swiss were able to persuade the Allies that they were the severely constrained type, agreement occurred at reservation price $\underline{a}$.

As a first assessment of our approach, predicted results correspond to actual ones. No firm agreement was concluded while the Allies were uncertain about the true type of Switzerland. Misperceptions and misunderstandings hindered any definitive conclusion. Once the Swiss were able to alter the Allies' beliefs, the minimally necessary conditions were immediately granted to Switzerland. I delay further interpretation of this first illustration to the conclusion, and move to my second empirical illustration.

4.2 EC and EFTA and the Creation of a European Economic Area

For this second empirical illustration, I shall focus on game specification and present only a quick appraisal of the bargaining.[51] I therefore demonstrate the structural correspondence between the model and the empirical situation, and neglect correspondence at the process level.

On 20 June 1990 the six member states of the European Free Trade Association (EFTA) and the twelve member states of the European Community (EC) launched formal negotiations over the creation of a European Economic Area (EEA). The main issues on the agenda were: (a) the legislative basis for achieving the free movement of goods, services, capital and persons; (b) the strengthening of co-operation in supporting policies, such as R&D, environment, education, working conditions and social welfare; and (c) the reduction of economic and social disparities between regions.

Who and under what conditions were the actors taking part in the negotiation process? Formally, the negotiations were multilateral between the twelve member states of the EC and the six member states of the EFTA (and later Liechtenstein). Practically, however, negotiations were bilateral between the Commission of the EC on the one hand, and the EFTA high-level negotiating group on the other. EFTA members had been forced to set up a common negotiating body to meet EC Commission President Jacques Delors's conditions for a broadening of co-operation between EFTA and EC.[52] The EC Commission spoke for the Community along the lines agreed with member states.

These two actors were markedly asymmetric. The EC was the result of the impetus of political, economic and cultural development in Western Europe, whereas the EFTA had experienced a continuing erosion and weakening of its position and lacked any project for the future. The respective interests in negotiation reflected such asymmetry. On the EFTA's side, all members valued highly a new scheme for co-operation with the EC. The main reason behind this interest was to be found in the economic dependence of these countries on the EC, both for exports and imports. Dependence was also social and political in a changing Europe. On the EC's side, stakes were clearly lower but far from negligible. Economically, EFTA countries were the largest trading partner, and the balance of trade was in favour of the EC. Politically, the new co-operation scheme with the EFTA would constitute one pillar of a larger European architecture designed to welcome East and Central European countries, through association treaties.[53] In addition, as suggested by Wallace and Wessels (1991,

p. 272), the EC could seek in the EEA a way to increase its influence on international regimes and world commercial negotiations.

Regarding the correspondence between the model and reality, the EFTA is actor A, eager to 'buy' closer association with the EC, actor B. The asymmetry in power is reflected in the structure of the game which favours actor B, which has a last say in the process and makes a 'take it or leave it' offer in the second period. The asymmetry in power is also reflected in the value of the discount factor. The EC countries were not time-impatient since they themselves set the future agenda. The EFTA was much more impatient, especially because member states wanted the future EEA to come into force from 1 January 1993.[54]

Ratification constraints shape the reservation value of each actor. Owing to the unanimity requirement on the EFTA's side and to the requirement of ratification by each member state on the EC's side, these constraints were especially important. Formally, therefore, reservation prices were dependent upon the domestic pivotal constituent of the most constrained member for the issue considered. Intraorganizational bargains, however, could be expected to reduce the salience of individual countries' domestic constraints and shift reservation prices closer to the position of the median price inside the EFTA high-level negotiating group and the EC Council, respectively.

Regarding the information on reservation prices, two options are plausible. The first option assumes that the effects of intraorganizational bargaining are negligible and considers the information to be complete on the EC's reservation price but incomplete on the EFTA's valuation. Ratification constraints in the EC members states were relatively easy to assess. Domestic constituents – national parliaments – did not care much about the EEA negotiations, except on some sensitive issues such as regional disparities, fishing rights and immigration policies. On these issues, the attitude of the different constituencies was clear-cut. On the EFTA's side, reservation values were hard to assess by the EC. This was especially true for Switzerland where ratification required a majority of both the people and the cantons. Information on the other countries' minimum acceptable prices was less problematic. In Sweden, Austria and Finland, domestic constraints were not binding for the issues on the agenda. Norway and Iceland were severely constrained but this mainly concerned aspects related to the fishing industry. In Norway qualified majority parliamentary approval (three-quarters), coupled with traditional 'Euro-scepticism', made any precise anticipation of the minimally acceptable deal difficult. The EC could therefore not be sure of the type of EFTA it would face, but could foresee two main alternatives: either the domestic constraints of the hard-liner countries were really as severe as they claimed (type $\underline{a}$ in the model) or they were less stringent and would not hinder a wider agreement (type $\bar{a}$).

Given these types, if we use the even distribution of beliefs as a good approximation of the puzzle for EC negotiators, the EC position could be seen as hawkish.

A far-reaching agreement with the loosely constrained type of EFTA only was preferred to a limited deal with both types of EFTA.

In the second option, intraorganizational bargaining significantly influences the minimal acceptable positions. Accordingly, both sides had an incomplete knowledge of the other's reservation price. Even if individual domestic ratification requirements for EC countries were relatively easy to assess, intra-EC deals could be expected to alter minimum acceptation levels. Depending on these internal dynamics, EFTA countries could face two possible types of EC: one severely constrained which would stand firm around the stance of southern members and Ireland, and one which would be more willing to compromise and was closer to the position of the big northern countries. These two types were not very different regarding their attitude under incomplete information. Both were hawkish and favoured a broad agreement with the loosely constrained EFTA only. On the other side, disagreement was too inefficient for type $\bar{a}$ whereas type $\underline{a}$ would forgo an agreement with type $\bar{b}$ and, rather, strike a deal with type $\underline{b}$. Accordingly, the loosely constrained EFTA was dovish, and the tightly constrained EFTA was hawkish.

Given these specifications, both options yield similar predictions. We can see from Propositions 2 and 3 that there is no equilibrium. The two types of EFTA do not pool and cannot be strictly separated. A whole range of outcomes is expected including delays in agreement and disagreement. Under complete information, an agreement is quickly reached and the exchange price is fixed at the EFTA's reservation price.

Are these rough predictions consistent with the empirical record? First, regarding delays, negotiations began in June 1990, but a general political agreement could not be reached before October 1991. In addition, unanticipated objection by the European Court of Justice postponed the official signing to May 1992. Switzerland, Iceland and Norway on the EFTA's side, Spain, Portugal, Greece and Ireland on the EC's side, hindered any agreement for a long time.

Secondly, demands for concessions often appealed to ratification requirements, especially on the EFTA's side. Domestic constraints were consciously used for strategic ends from the beginning of the process to its conclusion, and contributed to partial success for the EFTA countries. Special transition periods and specific permanent exceptions were granted to individual states in the most sensitive areas, such as Alpine transit, foreign immigration, capital investment and fishing rights.

The recourse to ratification hurdles might have brought better results if some strategic mistakes on the EFTA's side had been avoided. Indeed, over-exploitation of domestic constraints at the beginning of the process affected their subsequent credibility. On the other hand, intraorganizational pressures and unanimity requirements made a systematic use of domestic specificities politically impossible.

These limitations seriously affected Switzerland. Contrary to the case of the League of Nations, the Swiss government was trapped inside a collective organization and could not carry the exploitation of its domestic hurdles to its conclusion.[55] It could not assume alone a general breakdown of negotiations and had to give way under pressure from other EFTA members. However, it managed to get some individual concessions which were considered to be sufficient for a general agreement to belong to the feasible set. As the verdict of a referendum on 6 December 1992 would reveal, the Swiss executive probably misperceived the domestic pivotal constituent's position and thus was forced to an involuntary defection.[56]

5 CONCLUSION

There has been a growing interest in the analysis of domestic politics and international negotiations. Many scholars, however, have merely reproduced old results or fallacies and thus failed to bring new insights into the field. Moreover, research has failed to produce a modelling dialogue between formal developments on the one hand and careful empirical work on the other. This chapter has attempted to address these issues, and has helped to illuminate important elements regarding the correspondence between theoretical assumptions and the evidence presented in section 4.

Not only do the predictions derived from the model fit the actual outcomes, both for the timing and for the kind of agreement, but my analysis also helps to understand in depth the negotiation process.

For the case of the League of Nations first, our approach enables us to refine the Swiss executive's explanation that a combination of domestic specificities, international friendships and negotiators' strategic aptitude produced the actual outcome. Domestic specificities were the last recourse and finally secured an acceptable agreement. In their absence, it is highly doubtful that the Allies would have granted special status to Switzerland inside the League. The efficiency of such a bargaining device was, however, long affected by asymmetric information distribution among the actors. The Allies knew the existence of particular political ratification constraints inside Switzerland, but they were much more confident than the Swiss government about the positive attitude of the Swiss people regarding adhesion to the League of Nations. Their perception had to be altered and the Swiss negotiators benefited from international friendships to foster this change. Personal friendships proved more important than national ones in convincing the Allies' negotiators of the stringency of domestic hurdles. International sympathy had thus a limited influence and did not permit a stable agreement to be reached, mainly because of information problems.

Swiss negotiators skilfully combined domestic constraints and international friendships, giving support to our assumption of rational actors in situations of interdependence. They quickly understood that the key to their problems was to be found in the others' perceptions of Swiss domestic affairs, and their behaviour was continuously consistent with this awareness. The only exception was the December 1919 memorandum which was badly formulated and created the impression that the Swiss were demanding additional concessions. It was, however, quickly corrected with no consequences for the final deal.

Empirical evidence gives credit to the sequentiality of international and domestic processes as opposed to Putnam's simultaneous approach. The Swiss executive tried to restrict internal debate as much as it could before it obtained assurances at the international level. Serious and intense domestic discussions only began at the end of the international negotiations. The Swiss government did not try to influence the median voter position during the process, using neither the question of the headquarters of the League nor the economic implications of staying out of the League. The median voter position was considered as a fixed constraint and the negotiators did their best to come up with an acceptable arrangement. Similarly, Swiss constituencies were not strategically influenced by the Allies' governments or audiences.

Regarding negotiations over the European Economic Area, my quick assessment of the bargaining shows that domestic factors were indeed important factors in explaining the nature and timing of an agreement. The coalitional nature of the players, however, hindered full exploitation of domestic factors by the various players. As a predictable consequence, the strategic impact of domestic hurdles significantly decreased for countries like Switzerland.

The long duration of the process should not have come as a surprise, but many negotiators shared the contrary opinion at the beginning of the process, that an agreement could be concluded quickly. This inconsistency, coupled with inappropriate demands at the outset, raises questions about the strong rationality postulates regarding the behaviour of negotiators. However, the analysis of the last rounds of the process tends to show that negotiators did behave along the lines we assume in our model, and thus initial anomalies reflected strategic ineptitude. Negotiators neglected the crucial role of perceptions under incomplete information, and did not understand the far-reaching consequences of their behaviour.

Regarding domestic politics and international negotiations, empirical evidence supports my analytical view. Negotiating governments considered domestic constraints to be constant parameters and did not engage simultaneously in domestic and international games. They did not strategically consider the possible domestic repercussions of actions taken at the international level. Once the international process was over, they turned to ratification campaigns. These proved unproblematic except in Switzerland, where the international agreement failed to get domestic approval. Over-exploitation of domestic constraints was detrimental not only at the international level but also at the domestic level because it involuntarily raised popular expectations over a future agreement. The Swiss executive thus signed a treaty that it wrongly considered to be in its feasible set.

The central thrust of this chapter has been to foster a dialogue between sequential bargaining models and empirical work. The predictions and patterns of behaviour we derived from the model guided our empirical illustrations. In future work I shall go in this direction and extend empirical analysis based on a refined theoretical framework.

APPENDIX

Proof of Proposition 2

I do not offer a complete formal proof but sketch out the argument for the various ranges of parameters.

Case 1

I first look for pooling equilibria and then examine the possibility of separating equilibria (hybrid equilibria are only possible if we allow mixed-strategies).

Pooling equilibrium In equilibrium both types of A make the same first-period offer and thus actor B cannot update her beliefs. Accordingly, a dovish player B would prefer to offer $\underline{a}$, which guarantees her a payoff of $\delta_B(\underline{a} - b) > \delta_B(\bar{a} - b)/2$ which is the payoff she gets if she offers $\bar{a}$ in period 2. So in period 1 actor A offers p_1 such that $U_B(p_1,1) \geq \delta_B(\underline{a} - b)$ which gives $p_1 = \dot{v} = \delta_B\underline{a} + b(1 - \delta_B) \leq \underline{a}$. B's first-period strategy is to accept any offer greater or equal to $\dot{v}$ and reject any lower offer.

We now check that this constitutes a perfect Bayesian equilibrium with passive conjectures for out-of-equilibrium beliefs. Player B has no incentive to reject p_1 in period 1 because she is unwilling to ask $\bar{a}$ in period 2 and so would do at most equally in period 2. Player A has no incentive to offer more than $\dot{v}$ since he is sure that actor B will accept it. She has no incentive to offer $p_1 < \dot{v}$ either: since player B rejects such an offer and proposes $\bar{a}$ in period 2, type $\underline{a}$'s payoff is zero, which is clearly less than the equilibrium payoff; type $\bar{a}$ gets $\delta_A(\bar{a} - \underline{a}) \leq \bar{a} - \delta_B\underline{a} + b(1 - \delta_B)$, the latter being the payoff from offering $\dot{v}$ in period 1. No out-of-equilibrium offer can set in motion B's beliefs and then destroy this equilibrium.

Separating equilibrium In a strict separating equilibrium player B knows in period 2 what type of A he is facing and so the game is identical to the case under complete information. B offers $\underline{a}$ to type $\underline{a}$ and $\bar{a}$ to type $\bar{a}$. Going backward, we can infer B's behaviour in period 1. If he sees an offer made only by type $\underline{a}$ he accepts any offer greater than $\dot{v}$; if he sees an offer made only by type $\bar{a}$, he accepts any offer higher than $\ddot{v} = b(1 - \delta_b) + \delta_B \bar{a}$. Type $\bar{a}$ has a strong incentive to free-ride on type $\underline{a}$. This incentive is so strong that it is impossible to find a strict separating perfect Bayesian equilibrium. This is straightforward to see for any candidate equilibrium in which there is an agreement in period 1 with each type of A, but is more tedious for other cases. We only sketch the reasoning here for one specific system of beliefs and extrapolate from it.

Consider the case where B would get an agreement in period 1 only with type $\bar{a}$ with the following collection of strategies and system of beliefs. In period 1, type $\bar{a}$ offers $\ddot{v}$, type $\underline{a}$ offers anything lower than $\dot{v}$. B has the following conjectures:

$$\text{prob}(A = \underline{a}) = \begin{cases} 1 & \text{if } p_1 < \dot{v} \\ 0 & \text{if } p_1 \geq \dot{v} \end{cases} \tag{A7.1}$$

Accordingly, in period 1 B accepts any offer greater than $\ddot{v}$ and rejects any lower offer. In period 2, B offers $\underline{a}$ if he sees any offer less than $\dot{v}$ and offers $\bar{a}$ in case of any higher first-period offer. First, note that if

$$\delta_A > \frac{(\bar{a} - b)(1 - \delta_B)}{\bar{a} - \underline{a}},$$

this is not an equilibrium since type $\bar{a}$ has an incentive to deviate from his equilibrium behaviour. Secondly, this is a weak equilibrium since type $\underline{a}$ *is* completely indifferent between any offer equal to or lower than his reservation price. This has a crucial importance for B's out-of-equilibrium beliefs. Indeed, B knows that type $\underline{a}$ has a strong tendency to deviate from his strategy given the fact that he is indifferent, and his conjectures appear implausible. A similar reasoning for other potential equilibrium candidates shows that there is no strict separating equilibrium in pure strategies.

Case 2

Pooling equilibrium Since $(\underline{a} - b) < (\bar{a} - b)/2$, a hawkish actor B offers $\bar{a}$ in period 2 in any pooling equilibrium.

In period 1, A knows that B will reject for sure any offer lower than $\tilde{v} = \frac{1}{2}\delta_B \bar{a} + b(1 - \frac{1}{2}\delta_B)$. For a pooling equilibrium where both types of A offer $\tilde{v}$ in period

1, we must have $\tilde{v} \leq \underline{a}$, that is: $\delta_B \leq 2(\underline{a} - b)/(\bar{a} - b)$ (Δ). There exists a pooling PBE at $\tilde{v}$ iff (Δ) is fulfilled. To check that this is a PBE, we can proceed as for Case 1 (I skip this for reasons of parsimony).

Separating equilibrium Given our definition of the 'nature' of player B and given the fact that in a strict separating equilibrium player B has complete information in period 2, the reasoning presented under Case 1 applies equally well to Case 2. There is therefore no perfect Bayesian separating equilibrium.

Proof of Proposition 3

As for Proposition 2, the following is not a complete formal proof of Proposition 3, and the reasoning for separating equilibria applies to all cases even if I present it only under Case 1.

Case 1

Two types of B are dovish: we first examine the potential pooling equilibria, given passive conjectures for out-of-equilibrium actions, and then look for separating equilibria.

Pooling equilibrium There are two potential pooling equilibria $\check{v} = \underline{b}(1 - \delta_B) + \delta_B\underline{a}$, which is the minimal offer accepted by $\underline{b}$ in period 1, and $\hat{v} = \bar{b}(1 - \delta_B) + \delta_B\underline{a}$, which is the minimal offer accepted by both types of B in period 1.

1a: A hawkish type $\underline{a}$ always prefers to offer $\check{v}$ which is accepted only by type $\underline{b}$ because $\frac{1}{2}(\underline{a} - \check{v}) > (\underline{a} - \hat{v})$; a hawkish type $\bar{a}$ prefers $\check{v}$ when $\frac{1}{2}(\bar{a} - \check{v}) + \frac{1}{2}\delta_A(\bar{a} - \underline{a}) > (\bar{a} - \hat{v})$ equivalent to

$$\delta_A \geq \frac{(\bar{a} + \underline{b} - 2\bar{b}) + \delta_B(2\bar{b} - \underline{b} - \underline{a})}{\bar{a} - \underline{a}}.$$

Actor B has no interest in deviating from his strategy, given his beliefs (passive conjectures for out-of-equilibrium behaviour) and A's strategy.

1b: A dovish type $\underline{a}$ prefers $\hat{v}$ because $\frac{1}{2}(\underline{a} - \check{v}) < (\underline{a} - \hat{v})$ and a dovish type $\bar{a}$ always prefers $\hat{v}$ because

$$\frac{(\bar{a} + \underline{b} - 2\bar{b}) + \delta_B(2\bar{b} - \underline{b} - \underline{a})}{\bar{a} - \underline{a}} > 1$$

when both types of A are dovish.

1c: A hawkish type $\underline{a}$ always prefers $\check{v}$ (see Case 1a), and thus the only possible pooling offer is $\check{v}$. A dovish $\bar{a}$ has no incentive to deviate from this offer when

$$\delta_A \geq \frac{\left(\bar{a}+\underline{b}-2\bar{b}\right)+\delta_B\left(2\bar{b}-\underline{b}-\underline{a}\right)}{\bar{a}-\underline{a}}.$$

Contrary to case 1b, this condition can be satisfied for some values of $\delta_A \in [0,1]$.

Separating equilibrium The tricky part is to describe actor B's conjectures in period 1: when should he infer that A is dovish or hawkish, and can these conjectures make both types of A separate? Given the structure of the game, there can only be a separating equilibrium if $\bar{a}$ has an interest in signalling his type, given B's conjectures and subsequent strategies. If we use pessimistic conjectures, $\bar{a}$ has no incentive to make a separate offer, be she hawkish or dovish. But, let us consider other systems. First, with two hawkish A, assume that B's conjectures are the following:

$$\text{prob}(A=\underline{a}) = \begin{cases} 0 & \text{if } p_1 \geq \check{v} \\ 1 & \text{if } p_1 < \check{v} \end{cases} \tag{A7.2}$$

In period 1, type $\bar{a}$ offers $\underline{v} = \underline{b}(1-\delta_B) + \delta_B\bar{a}$, type $\underline{a}$ offers anything strictly smaller than $\check{v}$; type $\bar{b}$ accepts any offer $\geq \bar{v} = \bar{b}(1-\delta_B) + \delta_B\bar{a}$ and type $\underline{b}$ any offer $\geq \underline{v} = \underline{b}(1-\delta_B) + \delta_B\bar{a}$.

In period two, both types of B offer $\underline{a}$ if they see an offer $< \check{v}$, and $\bar{a}$ if they see any higher offer. This is a potential equilibrium if

$$\delta_A \leq \frac{(1-\delta_B)(\bar{a}-\underline{b})}{2(\bar{a}-\underline{a})}$$

and if $\delta_B > (\underline{a}-\underline{b})/(\bar{a}-\underline{b})$.

The problem, however, is that type $\underline{a}$ is indifferent between offering anything equal or lower than her valuation. In particular, she might systematically offer $\check{v}$ and this should set in motion B's beliefs, because he never sees any offer lower than $\check{v}$. He should then interpret this regular deviation as coming from type $\underline{a}$ and thus revise his beliefs accordingly. But this in turn will induce type $\bar{a}$ to offer $\check{v}$ in the first period since he would get more utility from his interaction with type $\underline{b}$ and at least the same from type $\bar{b}$.

Let us consider now a case with two dovish A's, the same conjectures for player B, and with the following collection of strategies: in period 1, type $\bar{a}$ offers

$\bar{v} = \bar{b}(1-\delta_B) + \delta_B\bar{a}$, a type $\underline{a}$ offers anything strictly smaller than $\hat{v}$; type $\bar{b}$ accepts any offer $\geq \bar{v}$ and type $\underline{b}$ any offer $\geq \underline{v} = \underline{b}(1 - \delta_B) + \delta_B\bar{a}$. In period two, both types of B offer $\underline{a}$ if they see an offer $< \hat{v}$, and $\bar{a}$ if they see any higher offer. This is a potential equilibrium if

$$\delta_A \leq \frac{(1-\delta_B)(\bar{a}-\bar{b})}{(\bar{a}-\underline{a})}$$

and if $\delta_B > (\underline{a} - \underline{b})/(\bar{a} - \underline{b})$. But we can rule out such an equilibrium by the same procedure as for the case for two hawkish A's.

Now let us consider a dovish type $\bar{a}$ and a hawkish type $\bar{a}$ with the following conjectures for player B:

$$\text{prob}(A = \underline{a}) = \begin{cases} 1 & \text{if } p_1 \leq \check{v} \\ 0 & \text{if } p_1 > \check{v} \end{cases} \qquad \text{(A7.3)}$$

Type $\bar{b}$ accepts any offer $\geq \bar{v}$, refuses any offer $p_1 \in [\hat{v}, \bar{v}]$, and rejects any offer $p_1 \in [\check{v}, \hat{v}]$. Type $\underline{b}$ accepts any offer $\geq \underline{v}$. In period two, both types of B offer $\underline{a}$ if they see an offer $p_1 \leq \check{v}$, and $\bar{a}$ if they see any higher offer. Here again the chosen set of conjectures does not appear to fit with the specificity of a separating equilibrium. If the price offered by type $\underline{a}$ is strictly separating, B should revise his beliefs differently and in particular should accept $\check{v}$.

Due to the structure of bargaining, there is no strong inefficiency in equilibrium and thus it is impossible to have the two types of A separate.

Case 2

Two types of B are hawkish: there are two potential *pooling equilibria*, $\check{w} = \underline{b}(1 - \frac{1}{2}\delta_B) + \delta_B\bar{a}$, which is accepted by type $\underline{b}$ only, and $\hat{w} = \bar{b}(1 - \frac{1}{2}\delta_B) + \delta_B\bar{a}$, which is the lower offer accepted by both types of B.

2a: Both types of A are dovish; in order to have a pooling equilibrium we have first to check whether type $\underline{a}$ can make such offers, that is whether $\underline{a} \geq \check{w}$ or $\underline{a} \geq \hat{w}$. Clearly, if we have $\check{w} > \underline{a}$, there is no possibility of a pooling equilibrium. For the case of two doves, however, there is no point in considering $\check{w}$, since both types of A clearly prefer to ask $\hat{w}$. To consider the possibility of a pooling at that offer, we must have the precondition $\underline{a} \geq \hat{w}$, or $\delta_B \leq 2(\underline{a} - \bar{b})/(\bar{a} - \bar{b})$. If this is satisfied it is straightforward to check that the following is a pooling equilibrium: both types of A offer $\hat{w}$, type $\underline{b}$ accepts any offer higher than $\check{w}$ and type $\bar{b}$ any offer higher than $\hat{w}$; in period 2, both types of B offer $\bar{a}$.

2b: If type $\underline{a}$ is dovish, and type $\bar{a}$ hawkish, we already know from Case 2a that type $\bar{a}$ cannot do better than offer $\hat{w}$ in a pooling equilibrium. Similarly, type $\underline{a}$

is better off with $\check{w}$, when he can make such an offer. If he cannot, then there is no possibility of having a pooling equilibrium. This implies that there is no equilibrium in this case.

Case 3

Type $\underline{b}$ is dovish, type $\bar{b}$ hawkish: we know from Cases 1 and 2 above that $\underline{b}$ will accept any offer greater than $\check{v}$ in period 1 whereas type $\bar{b}$ accepts any offer greater than $\hat{w}$. We can thus restrict the potential *pooling equilibria* to $\hat{w}$ and $\check{v}$. We first note that $\hat{w} > \check{v}$, given a hawkish type $\bar{b}$.

Let us first examine the conditions for having a pooling at $\check{v}$. For type $\bar{a}$ to stay at this equilibrium, we must have: $\frac{1}{2}(\bar{a} - \underline{b}(1 - \delta_B) - \delta_B\underline{a}) \geq \bar{a} - \bar{b}(1 - \frac{1}{2}\delta_B) - \frac{1}{2}\delta_B\bar{a}$ which leads to

$$\delta_B \leq \frac{2\bar{b} - \bar{a} - \underline{b}}{\bar{b} - \underline{b} - \bar{a} + \underline{a}} \quad \text{(A7.4)}$$

Condition (A7.4) can only be satisfied when type $\bar{a}$ is hawkish.

If type $\bar{a}$ has no incentive to deviate, neither has type $\underline{a}$ and thus we have a pooling at $\check{v}$ when both types of A are hawkish and (A7.4) is satisfied.

Let us now turn to a pooling equilibrium at $\hat{w}$. We first must have $\hat{w} \leq \underline{a}$ equivalent to

$$\delta_B \leq 2(\underline{a} - \bar{b})/\bar{a} - \bar{b} \quad \text{(A7.5)}$$

Type $\bar{a}$ has no incentive to deviate if (A7.4) is not true, which is surely verified if she is dovish. What can we say about type $\underline{a}$? For type $\underline{a}$ to stay on this equilibrium, we must have: $\frac{1}{2}(\underline{a} - \underline{b}(1 - \delta_B) - \delta_B\underline{a}) \leq \underline{a} - \bar{b}(1 - \frac{1}{2}\delta_B) - \frac{1}{2}\delta_B\bar{a}$ which leads to

$$\delta_B \geq \frac{2\bar{b} - \underline{b} - \underline{a}}{\bar{b} - \underline{b} - \bar{a} + \underline{a}} \quad \text{(A7.6)}$$

Condition (A7.6) cannot be satisfied if type $\underline{a}$ is hawkish and so we can sum up the situation in the following way:

3a: If both types of A are hawkish: we have a pooling equilibrium at $\check{v}$ if (A7.4) is satisfied; otherwise no equilibrium.

3b: If both types of A are dovish: if (A7.5) and (A7.6) are satisfied we have a pooling at $\hat{w}$, otherwise no pooling equilibrium exists.

3c: if type $\underline{a}$ is hawkish and type $\bar{a}$ dovish: no pooling equilibrium.

NOTES

1. In this chapter I shall use the terms 'negotiation' and 'bargaining' interchangeably. I follow the standard definition of Rubin and Brown (1975, p. 2): 'the process whereby two or more parties attempt to settle what each shall give and take, or perform and receive, in a transaction between them'.
2. An extensive review of the literature can be found in Dupont (1991) and in my forthcoming dissertation.
3. 'Audience' refers to any physical or psychological constraint that restrict the behaviour of a negotiator and to which he must account for his action (Rubin and Brown 1975, p. 43).
4. A systematic evaluation of the respective advantages and drawbacks of various modelling tools is developed in Dupont (1991).
5. Rubinstein's (1982) seminal paper develops an infinite horizon model under complete information. Rubinstein (1985) and Osborne and Rubinstein (1990) extend the analysis to incomplete information. Two-period sequential bargaining models with incomplete information have been studied by Cramton (1983), Fudenberg and Tirole (1983) and Sobel and Takahashi (1983). For an introductory survey of the game theoretic literature on bargaining, see Fudenberg and Tirole (1991), Kreps (1990) and Sutton (1986). Roth (1985) and Binmore and Dasgupta (1987) offer a collection of prominent contributions.
6. The analysis that follows shares some core characteristics and notation with Fudenberg and Tirole (1983). I am grateful to Robert Powell for pointing out to me this contribution and for the long discussions we had on it.
7. I treat player A as female and player B as male and thus I use different pronouns when referring to specific individuals.
8. Negotiations might, however, take place if both actors wrongly believe that $a > b$.
9. According to Black's median voter theorem, a median voter exists if the issue space is unidimensional and preferences are single-peaked.
10. For a discussion on the role of outside options, see Osborne and Rubinstein (1990, pp. 54–63); see also the models studied by Fudenberg, Levine and Tirole (1987) and by Perry (1986).
11. We restrict the possible offers to the zone of agreement and we do not allow for mixed strategies. If we envisage the negotiation as a single event with actors who will not interact under the same conditions in the future, mixed strategies have little empirical significance.
12. The standard 'accept if indifferent' assumption is used, following Fudenberg and Tirole (1983, fn 4) for its justification. If actor A refuses an offer $p_2 = a$, player B could change his offer a bit so that it will be accepted. With discrete types of actor A, player B has a strong interest to do so.
13. The generalization is straightforward but obscures the presentation without bringing in additional qualitative properties.
14. This definition only applies to cases of incomplete information and thus differs from those found, for example, in Aggarwal and Dupont (1992) and Bueno de Mesquita and Lalman (1992).
15. Our finding that there is no strictly separating equilibrium corresponds to that of Fudenberg and Tirole (1983) for a situation where the seller makes the first-period offers.
16. For wider treatments from the Swiss perspective, see in particular Ruffieux (1972), Stupan (1943), Rappard (1924); see also Soiron (1973) and Stettler (1969). For a specific historical treatment of the question of the headquarters, see Fleury (1981).
17. The Swiss people voted in favour of adhesion with a majority of 56.3 per cent, but only $11\frac{1}{2}$ cantons accepted it and $10\frac{1}{2}$ cantons rejected it (Stupan 1943, p. 159).
18. A status of integral neutrality would have exempted the Swiss from having to participate in any kind of collective sanctions, economic or military. Differentiated neutrality concerned military measures only.
19. 'Message complémentaire du Conseil Fédéral à l'Assemblée Fédérale concernant l'adhésion de la Suisse à la Société des Nations', *Feuille Fédérale*, (1920), 354.
20. Even if the League was all but an economic organization, the Federal Council thought that political isolation stemming from a non-participation would make trade relations more difficult and would

preclude many potential agreements. This argument was clearly stated in the 4 August 1919 message to the Parliament (*Feuille Fédérale*, **4** (1919), 671).

21. There was no scope for discussions on general matters regarding the Peace Conference, as President Wilson had already hinted to the Swiss 'unofficial' representative at the Paris Peace Conference, William Rappard, on 20 November 1918 (Rappard, 1956, pp. 55–6; quoted in Link, 1987c, pp. 626–9).
22. See Memorandum of Gustave Ador, President of the Swiss Confederation, on his conversation with Wilson in Paris, 23 January 1919 (*Documents diplomatiques suisses*, **7/1**, 246–8; trans. and reproduced in Link, 1987a, pp. 233–4).
23. See Memorandum of Gustave Ador, after his discussions in Paris on the eve of the Peace Conference (*Documents diplomatiques suisses*, **7/1**, 240–62). We shall see that, on the French side, later declarations were less friendly to Switzerland on the question of the headquarters.
24. The alternative is to consider that the value was much better known to the Allies because they were both willing and able to keep some information from the Swiss. We have not found evidence along these lines.
25. Small deviations from this configuration of beliefs do not significantly change the predicted patterns of behaviour.
26. The entry price would have surely been higher for any other neutral country, but Switzerland was considered as an exception.
27. This constraint was not mandatory according to the Constitution at that time, but the Federal Council had already chosen to make it mandatory for political reasons. See Rappard (1924, pp. 12, 29–30).
28. This became evident later in the process, especially during the debates in Parliament and the official campaign before the referendum.
29. Federal Councillor Calonder officially stated: 'We shall eventually be forced to ask the Swiss people . . . Regarding the conduct of negotiation, we are in a worse position than the others.' (quoted in Ruffieux, 1972, p. 65, my translation).
30. See the Memorandum Regarding the Neutrality of Switzerland, addressed to the Peace Conference on 8 February 1919, in *Documents diplomatiques suisses*, **7/1**, 352–4. Domestic specificities were never mentioned as a threat to a future agreement but only in the perspective to morally justify an exception to the question of neutrality. The appeal to American friendship had already been explicitly suggested in December 1918 by the special Advisor for Foreign Affairs, Lucien Cramer (*Documents diplomatiques suisses*, **7/1**, 118–21).
31. The special relationship with Wilson dates back to the time when Rappard was Assistant Professor of Political Science at Harvard while Wilson was Professor of Jurisprudence and Political Economy at Princeton.
32. See, for instance, the minutes of the ninth meeting of the Commission on 13 February 1919, in Miller (1928, p. 238). The delegates probably did not care because at that time they did not realize how tight a constraint it was to the Swiss government.
33. *Documents diplomatiques suisses*, **7/1**, 574–7.
34. For a detailed account of the discussions, see for instance the minutes of the 14th Meeting, in Link (1987b, pp. 222–6). For a detailed historical account of the intense diplomatic work around the question of the headquarters, see Fleury (1981) and Dupont (1992).
35. For a detailed account of the discussions and problems associated with the Swiss Amendment, see Miller (1928, pp. 428–38).
36. For a similar interpretation, see Ruffieux (1972, p. 73).
37. See Rappard's report in *Documents diplomatiques suisses*, **7/1**, 697–702.
38. See an account of it in Miller (1928, p. 436).
39. See Rappard's report in *Documents diplomatiques suisses*, **7/1**, 724
40. See *Feuille Fédérale*, **4** (1919), 567–81.
41. See Rappard's report on conversations with Cecil, Drummond and House in *Documents diplomatiques suisses*, **7/2**, 113–18; see also the despatch from the British Delegate in Bern, Sir Horace Rumbold, to Prime Minister Balfour in *Documents on British Foreign Policy, 1919–1939*, **5**, 228.
42. *Documents diplomatiques suisses*, **7/2**, 355.

43. This was labelled the American clause because ratification appeared to be seriously endangered only in the United States.
44. *Documents on British Foreign Policy*, **5**, 853, 890.
45. *Documents diplomatiques suisses*, **7/2**, 390–1.
46. *Documents on British Foreign Policy*, **5**, 654–5.
47. *Documents diplomatiques suisses*, **7/2**, 464–7.
48. *Documents on British Foreign Policy*, **2**, 930.
49. For accounts of these meetings, see the British minutes in *Documents on British Foreign Policy*, **2**, 928–32, 961–2; see also Ador's report in *Documents diplomatiques suisses*, **7/2**, 477–89.
50. *Documents diplomatiques suisses*, **7/2**, 509–10.
51. This is due to evident space limitation. A partial description of the negotiations can be found in Dupont (1992) and a full one will appear in my dissertation.
52. In a speech to the European Parliament on 17 January 1989 (see Agence Europe, doc. no 1542/1543, 26 January 1989).
53. See interview with Horst Krenzler, the EC chief negotiator, in *EFTA Bulletin*, **4/90**, 20.
54. *EFTA Bulletin*, **3/90**, 2.
55. See Dupont (1992) for a systematic comparative analysis of the two negotiations from a Swiss perspective.
56. The Swiss people rejected the EEA treaty with a majority of 50.3 per cent, and the cantons rejected it by a clear majority of 16 to 7.

REFERENCES

Aggarwal, Vinod (1989), 'Interpreting the History of Mexico's External Crises', in Barry Eichengreen and Peter Lindert (eds), *The International Debt Crisis in Historical Perspective*, Cambridge, Mass.: MIT Press, pp. 140–88.

Aggarwal, Vinod and Allan, Pierre (1992), 'Cold War Endgames', in Pierre Allan and Kjell Goldmann (eds), *The End of the Cold War: Evaluating Theories of International Relations*, Dordrecht: Martinus Nijhoff, pp. 24–54.

Aggarwal, Vinod and Dupont, Cédric (1992), 'Modeling International Debt Rescheduling: Choosing Game Theoretic Representations and Deriving Payoffs', paper presented at the American Political Science Association meetings, Chicago, September.

Binmore, Ken and Dasgupta, Partha (eds) (1987), *The Economics of Bargaining*, Oxford: Basil Blackwell.

Black, Duncan (1958), *The Theory of Committees and Elections*, Cambridge: Cambridge University Press.

Bueno de Mesquita, Bruce and Lalman, David (1992), *War and Reason*, New Haven, Conn.: Yale University Press.

Cramton, Peter C. (1983), *Bargaining with Incomplete Information: A Two-Period Model with Continuous Uncertainty*, Research Paper 652, Graduate School of Business, Stanford University.

Documents diplomatiques suisses, vols **7/1** and **7/2**, Bern: Benteli.

Documents on British Foreign Policy, 1919–1939, vols 2 and 5, London: His Majesty's Stationery Office.

Dupont, Cédric (1991), 'La négociation internationale: une analyse formelle des effets combinés de processus internes et externes', Geneva: Graduate Institute of International Studies, typescript.

—— (1992), 'Succès avec la SDN, échec avec l'EEE? Résistances internes et négociation internationale', *Annuaire suisse de science politique*, **32**, 249–72.

EFTA Bulletin, 1989–90.

Feuille Fédérale, 1919–20.

Fleury, Antoine (1981), 'L'enjeu du choix de Genève comme siège de la Société des Nations', in Saul Friedländer, Harish Kapur and Andre Reszler (eds), *L'Historien et les relations internationales,* Geneva: Institut universitaire de hautes études internationales, pp. 251–78.

Fudenberg, Drew and Tirole, Jean (1983), 'Sequential Bargaining under Incomplete Information', *Review of Economic Studies*, **50**, 221–47.

—— (1991), *Game Theory*, Cambridge, Mass.: MIT Press.

Fudenberg, Drew, Levine David K. and Tirole, Jean (1987), 'Incomplete Information Bargaining with Outside Opportunities', *Quarterly Journal of Economics*, **102**, 37–50.

Iida, Keisuke (1991), 'The Second Image under Uncertainty: a Game-theoretic analysis', Princeton, N.J.: Princeton University, typescript.

Iklé, Fred C. (1964), *How Nations Negotiate*, New York: Harper & Row.

Kreps, David M. (1990), *A Course in Microeconomic Theory*, Princeton, N.J.: Princeton University Press.

Link, Arthur S. (ed.) (1987a), *The Papers of Woodrow Wilson*, vol. 54, Princeton, N.J.: Princeton University Press.

——(1987b), *The Papers of Woodrow Wilson*, vol. 57, Princeton, N.J.: Princeton University Press.

—— (1987c), *The Papers of Woodrow Wilson*, vol. 63, Princeton, N.J.: Princeton University Press.

Meier, Heinz K. (1970), *Friendship under Stress: U.S.–Swiss Relations,1900–1950,* Bern: Herbert Lang.

Miller, David Hunter (1928), *The Drafting of the Covenant,* vol. I, New York: G.P. Putnam's Sons.

Myerson, Roger B. (1992), 'On the Value of Game Theory in Social Science', *Rationality and Society*, **4**, 62–73.

Osborne, Martin J. and Rubinstein, Ariel (1990), *Bargaining and Markets*, San Diego, Cal.: Academic Press.

Perry, Motty (1986), 'An Example of Price Formation in Bilateral Situations: a Bargaining Model with Incomplete Information', *Econometrica,* **54**, 313–21.

Putnam, Robert D. (1988), 'Diplomacy and Domestic Politics: the Logic of Two-level Games', *International Organization*, 42, 427–60.

Raiffa Howard (1982), *The Art and Science of Negotiation,* Cambridge, Mass.: Belknap Press.

Rappard, William (1924), *L'Entrée de la Suisse dans la Société des Nations*, Geneva: Sonor.

—— (1956), 'Woodrow Wilson, la Suisse et Genève', in *Centenaire Woodrow Wilson*, Geneva: Atar, pp. 29–74.

Roth, Alvin E. (ed.) (1985), *Game-theoretic Models of Bargaining*, New York: Cambridge University Press.

Rubin, Jeffrey A. and Brown, Bert R. (1975), *The Social Psychology of Bargaining and Negotiations*, New York: Academic Press.

Rubinstein, Ariel (1982), 'Perfect Equilibrium in a Bargaining Model', *Econometrica*, **50**, 97–109.

—— (1985), 'A Bargaining Model with Incomplete Information about Time Preferences' *Econometrica*, **53**, 1151–72.

Ruffieux, Roland (1972), 'L'Entrée de la Suisse dans la Société des Nations (1918–1920)', in Roland Ruffieux (ed.), *La Démocratie référendaire en Suisse au XXème siècle*, Fribourg: Ed. de l'Université, pp. 39–118.

Scharpf, Fritz W. (1991), 'Games Real Actors Could Play: the Challenge of Complexity', *Journal of Theoretical Politics*, **3**, 277–304.
Schelling, Thomas C. (1956), 'An Essay on Bargaining', *American Economic Review*, **46**, 281–306.
—— (1960), *The Strategy of Conflict*, Cambridge, Mass.: Harvard University Press.
Sobel, Joe and Takahashi, Ichiro (1983), 'A Multistage Model of Bargaining', *Journal of Economic Studies*, **50**, 411–26.
Soiron, Rolf (1973), *Der Beitrag der Schweizer Aussenpolitik zum Problem der Friedensorganisation am Ende des Ersten Weltkrieges*, Basel: Helbing & Lichtenhahn.
Stettler, Peter (1969), *Das aussenpolitische Bewusstsein in der Schweiz (1920–1930)*, Zurich: Leemann.
Stupan, Sergio (1943), *Comment la Suisse a adhéré au Pacte de la Société des Nations*, Lausanne: Imprimerie Centrale.
Sutton, John (1986), 'Non-cooperative Bargaining Theory: an Introduction', *Review of Economic Studies*, **53**, 709–24.
Wallace, Helen and Wessels, Wolfgang (1991), 'Conclusions', in Helen Wallace (ed.), *The Wider Western Europe*, London: Frances Pinter, pp. 268–81.
Walton, Richard and McKersie, Robert (1965), *A Behavioral Theory of Labor Negotiations*, New York: McGraw-Hill.

8. Some game-theoretical considerations on negotiations about migrations

Urs Luterbacher and Jean-Paul Theler

Over the last decade, the world has seen a tremendous increase in the magnitude of interregional and international migrations. Some of these are due to an increase in the political instability of various regions of the world and the numerous conflicts, civil, interethnic and international, that result from these difficulties. It is easy to see that migratory movements which are a consequence of political instabilities can then in turn generate more political problems: an important influx of migrants into a particular region will generate resentment on the part of the local population. This was demonstrated in the case of the Bangladeshis who fled to India as a result of the 1975 war between India and Pakistan thereby creating political turmoil in Assam, where the local population did not like this intrusion.[1]

However, other factors also play a role in migrations, such as income disparities between North and South, different demographic evolutions within these two regions of the world, and last but not least, a significant reduction in real transportation costs. This latter factor, while not at the origin of a decision to migrate, makes such a choice much easier to make. In addition to these more traditional causes of population movements, if projected global climate changes occur for whatever reasons, a new category of migrants, so-called environmental refugees, could be added to the list and contribute to the already substantial political problems raised by migrations. It has to be realized that, especially in this case, pull factors such as wage or standard of living differentials between North and South account for only part of the picture. Social systems under economic or environmental stress also have ways of pushing people away from them. Such push factors can even be enhanced by governmental policies that encourage migratory movements directly or indirectly. For instance, by providing incentives to move to other, maybe rural, regions, a government could ultimately provide the impetus for people to go further.

One would therefore face here some of the political consequences of global environmental change in the form of global migration. These would concern distribution problems posed by a likely increase in the flow of people from developing to industrialized countries. These problems are already acute now and could become more so if the climate evolves unfavourably in marginal production areas or if important coastal regions are flooded. A model of a marginal regional system developed recently[2] shows that such a society of small agricultural producers is, mostly for economic reasons, prone to 'demographic collapse', that is, massive emigration that reduces its population size to a bare minimum. Such a demographic collapse can be accelerated by adverse weather conditions that reduce the long-term productivity of marginal agriculture. If many such collapses occur, the population of more advanced economic areas in developing countries, particularly their cities, will increase and direct emigration to industrialized countries will also occur.

To understand the distributive problem created by the movement of populations and their consequences on the local labour markets, a game-theoretical model that deals with government decision-making about migrations can be set up. Our model will not deal with the problem of how decisions to migrate are made by individuals and then aggregated, or how such movements are accelerated by local social systems or by governments, but will analyse governmental decisions to restrict migrations and look at possible tradeoffs in this respect.[3] As seems obvious from our example of Bangladeshis moving to India, or from numerous instances of foreigner- or refugee-bashing in various parts of the world and of the success of some anti-foreign parties in national or local elections in Western Europe, governments will be pressured to limit free access to their territories. In fact, most governments will attempt to control and usually severely restrict the quantities of foreigners allowed to come into the country and participate in the domestic labour market. However, such policies will be difficult to implement in democratic societies without impeding excessively the freedom of movement of individuals. Such policies will also interfere with tourism and business activities which are economically profitable. In addition, as long as controls exist that tend to isolate the domestic labour market from the rest of the world, incentives will be provided either to attract illegal labour or for foreigners to sneak illegally into a country in the hope of finding better conditions. Therefore, governments can only hope to restrict, and not to stop, migration. The extent of such restrictions and their relations with economic factors and incentives have to be examined. Standard economic analysis tells us that movements of people can be substituted to some extent by movements of goods and services, and other factors of production.[4] Governments of industrialized countries have also a small margin to manoeuvre in this respect since protectionist policies towards domestic agriculture severely limit possibilities of increasing significantly the flow of goods from developing to more advanced economies. Therefore, the movement of other factors of production such as capital represents practically the only possibility of substitution to the movement of people. This particular aspect will be emphasized in our model.

To show how a game-theoretical model dealing with this question can be constructed, we assume a simple case of two countries, one from the industrialized North and the other from the developing South. Each country will be described in terms of a social welfare or utility function. Such a description raises the question of the accuracy of such social welfare functions in terms of the aggregation of individual preference functions. To make things easy at this point, we shall assume that they represent at least government preferences.[5] We shall now proceed with the description of a model for each country.[6] The following represents the North's welfare function:

$$U_N = F_1(R_N, E)$$

where R_N stands for revenue of a country of the North, E for a ratio of foreign labour L_{SN} over domestic labour L_{NN} which the government of the North seeks to control and U_N for the social welfare function. The ratio E is thus simply defined as:

$$E = L_{SN}/L_{NN}$$

We shall assume further that:

$$\partial U_N/\partial R > 0$$

and:

$$\partial U_N/\partial E < 0$$

In addition, we shall assume that second cross-derivatives vanish, which means that the welfare function is separable into its two arguments of R and E.

The relations defined above (first derivative of U_N with respect to E negative), imply that the government of the North is reluctant to accept foreign labour because of opposition from the population; this could have electoral and other consequences, such as political instability.

We shall also have the GNP of N which is:

$$Y_N = F_2(K_N, L_N, T_N)$$

$$\partial Y_N/\partial K_N > 0 \quad \partial Y_N/\partial L_N > 0$$

$$\partial Y_N/\partial T_N > 0$$

where F_2 is a Cobb–Douglas-type production function and K_N, L_N, T_N stand for labour capital and technology of the North. All partials of F_2 are supposed to be positive. But the second derivatives are assumed to be negative, and again, since we assume that the production function can be separated into its arguments, the second cross-derivatives are assumed to be zero. We shall also assume that:

$$R_N = Y_N - a(\partial Y_N/\partial L_N)E\,L_{NN} + r(\partial Y_S/\partial K_S)K_R\,K_{SS}.$$

In other words, the revenue of the North depends on GNP minus the remittances of foreign workers. We assume here that northern labour is paid according to its marginal productivity and that there is a situation of equilibrium in the labour market; that is, labour supply in the North equals labour demand (no unemployment). It is also assumed that the North exports capital to the South for which it gets a revenue proportional to capital imports of the South MK. The South will allow a certain ratio $KR = MK/K_{SS}$ and no more to get into the country. The parameters a and r represent the fact that not all wages of foreign workers or, respectively, revenues of capital are repatriated. Taxation and other expenses tend to limit the quantities brought home to ratios defined by a and r. In addition we can define the following:

$$\text{Total labour force, North: } L_N = L_{SN} + L_{NN}$$

Where L_{SN} is the migrant labour from the South, and L_{NN} the domestic labour force of the North.

Technology in the North is an increasing function of time, t, and also of capital imports into the South, because new knowledge can be gained from exporting capital to other countries:

$$T_N = F_3(MK,t).$$

Domestic capital stock in the North is assumed to be exogenous. Thus:

$$K_N = \text{EXOG}.$$

MK, capital exported to the South, is also thought to be exogenous.

The South has also a welfare function:

$$U_S = G_1(R_S,un,KR)$$

where in addition to revenue, R_S, unemployment, un, and KR, the foreign to domestic capital ratio, plays a role. We shall assume here that too much capital influx and too much unemployment is disliked, thus leading to:

$$\partial U_S/\partial R_S > 0,\ \partial U_S/\partial un < 0,\ \partial U_S/\partial KR < 0.$$

We can now define the GNP of the South in a similar way to that of the North:

$$Y_S = G_2(K_S, L_S, T_S) \text{ All partials} > 0$$

$$K_S = K_{SS} + MK \text{ where } K_{SS} = \text{Domestic capital stock of the South}$$

and the revenue:

$$R_S = Y_S + a(\partial Y_N/\partial L_N)L_{SN} - r(\partial Y_S/\partial K_S)KR\ K_{SS}.$$

In the South, then, the labour force that migrates to the North brings in additional revenues in the form of remittances. However, revenues from imported capital KM go back to the North, at least in part.

Here the foreign to domestic capital ratio KR is defined as $KR = MK/K_{SS}$. Southern technology benefits from Northern capital and it is also represented as an increasing function of time:

$$T_S = G_3(MK, t).$$

The domestic Southern capital stock K_{SS} is also thought to be exogenous.

The nature of the labour force in the South deserves special discussion. Since the South exports labour, it experiences unemployment conditions. Presumably such unemployment is generated, because in many cases the wages associated with the marginal productivity of labour are below subsistence level. Other rigidities might also exist, such as control of the labour market by specific organizations. We shall therefore assume the existence of a minimal wage level $w(0)$. The labour demand schedule in the South, L_S, will thus be determined by the ratio of the marginal productivity of labour over the minimal wage level $w(0)$:

$$L_S = G_4\left(\frac{\partial Y_S / \partial L_S}{w(0)}\right).$$

The unemployment level in the South, un, can now be determined on the basis of the difference between the potential labour force given by the demographic structure of the South, L_D, and the labour demand schedule L_S:

$$un = \frac{L_D - L_S - L_{SN}}{L_D}$$

This completes the presentation of the model, which can now be analysed in terms of its major characteristics in order to determine the nature of the game-theoretical equilibria it generates.

DISCUSSION

Two basic situations will be analysed in the context of the model presented above:
1. A static one where governments of both North and South decide on the basis of their immediate interests without trying to anticipate each others' responses.
2. A dynamic perceptive where both governments have a longer term vision of their own interests and seek to respond to the other side's changes in strategy.

Static Situation

We can show (the proof is presented in Appendix I) that the partials of the GNP and revenue of the North with respect to L_{SN} are positive; so are the partials of GNP and revenue of the South with respect to *MK*. In the South, in addition to the revenue effect of capital on income (GNP) and thus on global revenue, additional capital brought by the North has the advantage of reducing *un* (the unemployment rate of the South) as additional capital raises the marginal productivity of labour and thus the wage rate.[7] Both countries would thus have an incentive to open up their borders to some extent. This could be done in the form of an agreement where admission of more foreign workers by the North would be exchanged for the admission of more capital from the North by the South. However, given the negative impacts of (a) the foreign worker ratio on the social welfare function of the North, and (b) the foreign capital ratio on the social welfare function of the South, maximizing U_N and U_S with respect to *E* and *KR* respectively without an agreement will bring about relatively low levels of *E* and *KR* (proof is given in Appendix I under the assumption of negative partial derivatives of both *E* and *KR* in each welfare function and some additional minor assumptions). If, however, there is an agreement for more openness on the part of both North and South to let more labour or, respectively, more capital in, then both regions are better off in terms of their respective social welfare functions up to the point where additional labour or capital produces diminishing returns in terms of the production functions. We assume here, however, that we are well below such levels. If this latter condition obtains, we are therefore in a Prisoner's Dilemma situation where mutual agreement increases the welfare functions of both but where there is an incentive to cheat on the agreement because unilateral gains will be high in such a case. The proof that the two countries are indeed in such a Prisoner's Dilemma situation is given in Appendix I. These two regions are thus reluctant to negotiate such an agreement in the short term since a stable Nash equilibrium exists when each of them chooses to close its borders as much as possible to the other's factors of production. Does a longer-term perspective lead to similar conclusions? To answer such a question requires an analysis of a more dynamic situation.

Dynamic Situation

A way out of the above-mentioned Prisoner's Dilemma is to take a long-term view with a specific way of discounting the future. In this case, each region would maximize its specific welfare function over time, subject to a particular discount rate. For instance, a way of representing such a situation would be for each country to maximize a particular integral of its welfare function:

$$\mathrm{Max}(\int\left(U_N e^{-\theta_1(t-\tau)}\right)d\tau;\ \mathrm{Max}(\int\left(U_S e^{-\theta_2(t-\tau)}\right)d\tau$$

where θ_1 and θ_2 represent the discount rates. This maximization would be subject to the constraint that each country would observe the other in terms of its foreign worker or foreign capital ratio, respectively, and adjust its policy immediately to that of the other country. In other words, each country's government would be able to react almost immediately to a change in the admission policy of the other, leaving open the possibility of retaliation. Such observation and retaliation processes could be expressed by the following differential equations:

$$\frac{\mathrm{d}E}{\mathrm{d}t} = \varphi KR;\ \frac{\mathrm{d}KR}{\mathrm{d}T} = \psi E$$

where φ and ψ represent adjustments countries would make to their respective increase or decrease in allowed foreign worker or foreign capital ratios as function of the others' policies. Maximizing the above integrals with the constraints expressed in terms of the differential equations representing mutual attention to each others' actions amounts to a dual optimal control or differential game problem. In this case, the differential game is of the so-called bang-bang variety, meaning that the control variables φ and ψ will (optimally) switch abruptly from one value to another, here a maximum, a minimum or 0.[8] If we hypothesize that they will fluctuate within some reasonable interval, we shall be interested in conditions that drive them both to their maximum. Such conditions are achieved for the North, as we can show (the proof is in Appendix II), if the productivity of capital is sufficiently high in the South to make up for the negative effects of an additional influx of foreign workers on the population of the North. Similarly, for the South, the revenues from guest-workers and the reduction of unemployment have to compensate for the negative impact of additional foreign capital. A similar reasoning holds for a discrete time-differential game approach. It is interesting to notice in this context that our proof is independent of the existence of a particular discount rate. Moreover, we also show that such policies can be implemented without having complete information about the other side's retaliation policies expressed by the values of φ and ψ. A *tâtonnement* sequence will converge to the estimation of the correct values of φ and ψ on both sides, that is, a situation where $\varphi = \psi$, as the proof in Appendix II shows.

Our approach allows also, therefore, for the introduction of incomplete information problems. Countries can try to hide their true goals, in terms of foreign worker and capital ratios. However, the above-mentioned *tâtonnement* sequence will inevitably lead to adjustments that will force them to reveal their true preferences.

Other distribution problems can be set up in a similar way; in particular, those dealing with possible and differential rates of reduction of revenues due to trade or anti-emission policies. Thus our general framework makes widely accessible a whole range of political problems connected to global bargaining and global change.

APPENDIX I

We wish to demonstrate that the bargaining situation envisaged in this Chapter takes the form of a Prisoner's Dilemma, as shown in the following proposition:

> *Proposition I.1*: The respective preference structures of the authorities of the North and of the South, concerning the foreign labour and foreign capital ratios E that they are willing to tolerate, leads to a Prisoner's Dilemma type game.
>
> *Proof*: To prove the above proposition it is sufficient to show that a game solution resulting from an agreement to trade off foreign capital for foreign labour and vice versa is Pareto-superior to a Nash equilibrium that would derive from a mutual maximization of the utility functions of the North and the South without regard to the other side. If such a Pareto-superior situation exists, it is easy to see that each country has an incentive to cheat on the agreement. Cheating by not allowing in as much foreign labour or foreign capital as agreed will leave one side better off than in the Pareto-optimal situation (of the agreement) while leaving the other side worse off than in the Nash equilibrium if it abides by the agreement.
>
> It is therefore sufficient to show that for both North and South:

$$\underset{L_{SN}/MK}{\text{Max}} (U_{N/S} \mid \text{Agreement } E = \phi\, KR \text{ or } KR = \psi E) > \underset{L_{SN}/MK}{\text{Max}} (U_{N/S} \mid \text{No Agreement.})$$

and this everywhere.

It is enough to show this for one country; the demonstration can then be symmetrically applied to the other. For the North, the above inequality holds: maximization of U_N with respect to E amounts to the same as maximization with respect to L_{SN}, since the Northern labour force is taken as exogenous. The same can be said about the maximization of KR by the South. Maximization with respect to MK amounts to the same. Maximization of U_N leads to the following expression:[9]

$$\frac{\partial U_N}{\partial L_{SN}} = \frac{\partial U_N}{\partial R_N}\left[(1-a)\frac{\partial Y_N}{\partial L_N} - aL_{SN}\frac{\partial^2 Y_N}{\partial L_N^2}\right] + \frac{1}{L_{NN}}\frac{\partial U_N}{\partial E} = 0$$

The first term in the above equation is positive if one assumes that $\partial^2 Y_N/\partial L^2{}_N$ is < 0 while the second term is always negative. The solution of the above equation in terms of L_{SN}, say $L_{SN}{}^*$, gives the optimal level of foreign labour that will be allowed into the North (second-order conditions lead to a maximum if one assumes $a < 50\%$). A similar reasoning will give us an optimal MK^* solution for the South. The pair $L_{SN}{}^*$, MK^*, with utility levels $U_N{}^*$, $U_S{}^*$ will bring about a Nash equilibrium since there is no incentive for the two country authorities to change strategies unilaterally.

Making an agreement $MK = \psi L_{SN}$ leads to the following maximization condition:

$$\frac{\partial U_N}{\partial R_N}\left[(1-a)\frac{\partial Y_N}{\partial L_N} + \psi\frac{\partial Y_N \partial T_N}{\partial T_N \partial MK} - a\, L_{SN}\frac{\partial^2 Y_S}{\partial L_N^2} + r\psi\frac{\partial Y_S}{\partial K_S}\right] +$$

$$\frac{1}{L_{NN}}\frac{\partial U_N}{\partial E} + \psi\frac{\partial^2 Y_N}{\partial K_S^2} r\ KR\ K_{SS} = 0$$

Subtracting the expression calculated with the assumption of no agreement from the expression above we get:

$$\frac{\partial U_N}{\partial R_N}\left[\psi\frac{\partial Y_N}{\partial T_N}\frac{\partial T_N}{\partial L_{SN}} + r\psi\frac{\partial Y_S}{\partial L_S}\right] + \psi\frac{\partial^2 Y_S}{\partial K_S^2} r\ KR\ K_{SS} > 0$$

which is positive everywhere if we assume reasonably that the second partial derivative on income Y_S with respect to the total capital stock of the South is close to zero. This assumption reflects the fact that the South needs a lot of capital and that diminishing return situations have not been reached yet. If relation (24) is positive everywhere for the reasons outlined above, then equation (23) leads also to a higher foreign labour level L_{SN}^{**} than L_{SN}^{*}, the solution of equation (22). This follows from the fact that both left hand sides of equations (22) and (23) are decreasing functions of L_{SN}. Since the left hand side of (23) is always above (22) in terms of L_{SN} as established by relation (24), the solution of equation (23) is also higher than the solution of equation (22). This leads then to a higher utility level U_N^{**} for the following reasons: from the above expression (24) we know that the partial derivative of U_N with respect to L_{SN} is greater everywhere under the condition of an agreement as opposed to the condition of no agreement. Since the solution of equation (23), L_{SN}^{**} is greater than the solution of equation (22) L_{SN}^{*} the utility level U_N^{**}, due to L_{SN}^{**} is also greater than the utility level U_N^{*} due to L_{SN}^{*}. The reasoning is symmetrical for the South where we get a similar higher utility level U_S^{**} with respect to the agreement capital level MK^{**}. From this we can conclude that the utility pair $U_N^{**}, U_S^{**} > U_N^{*}, U_S^{*}$. The utility pair due to the agreement is therefore Pareto superior to the one obtained without agreement which completes our proof. QED

APPENDIX II

We wish to maximize

$$\int_0^\infty U_N e^{-\theta_1(t-\tau)} \mathrm{d}t$$

for the North and

$$\int_0^\infty U_S e^{-\theta_2(t-\tau)} \mathrm{d}t$$

for the South.

Subject to the constraint for the North:

$$\frac{\mathrm{d}E}{\mathrm{d}t} = \varphi KR$$

and for the South:

$$\frac{dKR}{dt} = \psi E$$

this is a bang-bang problem where φ and ψ will have either

values depending on whether the sum of the solutions for the costate differential equations in λ are > 0 (max) or < 0 (min).

More precisely the function $\alpha_i(t)$ below has to be greater than 0 for φ and ψ to be at their maximum values. If $\alpha_i(t)$ is smaller than 0 then φ and ψ take minimum values. If $\alpha_i(t) = 0$, then φ and ψ will take arbitrary values:

$$\alpha_i(t) = \sum_{k=1}^{n} \lambda_k b_{kt}(x,t)$$

where the vector $b_{kt}(x,t)$ represents the right-hand side of the constraint equations, that is, here, φKR and ψE. Since these are positive by definition for positive φ and ψ we only need to look at the sign of the sum of the λ solutions.

The Hamiltonian for our problem can be written as:

$$\left[(\lambda_1 \varphi KR + \lambda_2 \psi E)\right] - U_N e^{\theta_1 \tau}$$

for the North and

$$\left[(\lambda'_1\, \varphi KR + \lambda'_2\, \psi E)\right] - U_S e^{\theta_2 \tau}$$

for the South.

The following costate equation system can thus be derived:

$$\dot{\lambda}_1 = -\psi\lambda_2 + \partial U_N / \partial E\, e^{\theta_1\tau}$$

$$\dot{\lambda}_2 = -\varphi\lambda_1 + \partial U_N / \partial KR\, e^{\theta_1\tau}$$

for the North and

$$\dot{\lambda}'_1 = -\psi\lambda'_2 + \partial U_S / \partial E\, e^{\theta_2\tau}$$

$$\dot{\lambda}'_2 = -\varphi\lambda_1 + \partial U_S / \partial KR\, e^{\theta_2\tau}$$

for the South.

This leads us to prove the following *sufficient* condition for maximization of the two overtime payoff functions in such a way that φ and ψ take maximum values, that is, that openness is assured at the highest possible level.

Proposition II.1: Both countries will spontaneously set φ and ψ at their highest value and reach a Nash equilibrium if the following conditions obtain:

(a) For both countries, the absolute values of the marginal utilities of the foreign labour ratio $\partial U_{N/S}/\partial E$ and of the foreign capital ratio $\partial U_{N/S}/\partial KR$ are equal.

(b) For both countries the decision made by the other to open borders has to be greater than or equal to his own, which implies $\varphi \geq \psi$ for the North and $\psi \geq \varphi$ for the South, which implies $\varphi = \psi$. This particular condition suggests a bargaining mechanism that can be envisaged to equalize φ and ψ if they are not the same. The proof for the convergence of φ and ψ is the following: if φ and ψ are different, then E and KR will be different. If these are different, U_N and U_S, respectively, will be lower, since a higher increase in E or, respectively, KR will not be compensated by a same increase in the other variable. Each country's ruler will therefore have an incentive to adjust the values of φ and ψ until they are equal. This mechanism suggests also, the existence of an equilibrium under conditions of incomplete information.

Proof: The evaluation of the solution of the costate equation system for the North leads to the following steps:

(a) Using the D (= d/dt) operators and solving for λ_1, λ_2 we get:

$$\left(D^2 - \psi\varphi\right)\lambda_1 = D\left(\partial U_N e^{\theta_1 \tau} / \partial E\right) - \psi \partial U_N e^{\theta_1 \tau} / \partial KR$$

$$\left(D^2 - \psi\varphi\right)\lambda_2 = D\left(\partial U_N e^{\theta_1 t} / \partial KR\right) - \phi \partial U_N e^{\theta_1 t} / \partial E$$

(b) Solutions for $\lambda_1(\tau)$ and $\lambda_2(\tau)$ lead to the following expressions:

$$\lambda_1(t) = C_1 e^{\sqrt{\psi\varphi t}} + C_2 e^{-\sqrt{\psi\varphi t}} + \int\left[e^{\sqrt{\psi\varphi(t-\tau)}} - e^{-\sqrt{\psi\varphi(t-\tau)}}\right] \left[d / d\tau\left(\partial U_N / \partial E\, e^{\theta_1 \tau}\right) - \psi \partial U_N\, e^{\theta_1 \tau} / \partial KR\right] d\tau$$

$$\lambda_2(t) = C_3 e^{\sqrt{\psi\varphi t}} + C_4 e^{-\sqrt{\psi\varphi t}} + \int\left[e^{\sqrt{\psi\varphi(t-\tau)}} - e^{-\sqrt{\psi\varphi(t-\tau)}}\right] \left[d / d\tau\left(\partial U_N / \partial KR\, e^{\theta_1 \tau}\right) - \varphi \partial U_N\, e^{\theta_1 \tau} / \partial E\right] d\tau$$

(c) If we take the sum $\lambda_1(t) + \lambda_2(t)$ we note that $\lambda_2(t)$ is always positive (assuming C_3 and $C_4 > 0$). The terms under the integral sign are always positive since $\partial U_N/\partial KR > 0$ and $\partial U_N/\partial E < 0$. $\lambda_1(7)$ on the other hand is either > or < 0 since the terms under the integral sign are always < 0. Since θ_1, the discount coefficient, is the same for λ_1 and λ_2, all the terms under the integral signs with the exception of the time derivatives will vanish when λ_1 and λ_2 are added and if the conditions enumerated above hold ($\partial U_N/\partial E = \partial U_N/\partial KR$, $\varphi = \psi$).

(d) The two time derivatives which are of opposite sign will also vanish for the following reasons: if we take them separately as:

$$\int\left[e^{\sqrt{\psi\phi(t-\tau)}} - e^{-\sqrt{\psi\phi(t-\tau)}}\right]\left[d / d\tau\left(\partial U_N / \partial E\, e^{\theta_1 \tau}\right)\right] d\tau$$

and

$$\int\left[e^{\sqrt{\psi\phi(t-\tau)}} - e^{-\sqrt{\psi\phi(t-\tau)}}\right]\left[d / d\tau\left(\partial U_N / \partial KR\, e^{\theta_1 \tau}\right)\right] d\tau$$

we can use a well-known theorem on integrals which states that, provided $\partial U_N/\partial E\, e^{\theta_1 \tau}$ and $\partial U_N/\partial KR\, e^{\theta_1 \tau}$, respectively, are of exponential order, that is, $\leq Me^{\alpha\tau}$ for some M and some α, than an integral

$$\int_0^\infty e^{-st} f'(t)dt$$

can be written as

$$s\int_0^\infty e^{-st} f(t)d - f(0).$$

Here, of course, $f(t)$ would be either $\partial U_N/\partial E\, e^{\theta_1 t}$ or $\partial U_N/\partial KR\, e^{\theta_1 t}$.

(e) If we use the above transformation, given that $\partial U_N/\partial E$ and $\partial U_N/\partial KR$ are of opposite sign and equal and that at $t = 0$, $e^{\theta_1 t} = 1$, all the terms under the integral sign vanish when $\lambda_1(t)$ and $\lambda_2(t)$ are added. This leaves only positive terms. Thus φ and ψ are at a maximum value. since a relation is made between E and KR with max φ and ψ, openness is guaranteed and each country gets more than it gets in isolation. This completes the proof for the North.

For the South, the procedure is essentially symmetrical. The proof is thus also valid for the South. Obviously, the positions of openness which both North and South reach in this process represent a Nash equilibrium, since no country authority has an incentive to deviate from its policies which correspond to optimal strategies. QED.

NOTES

1. This example is discussed in T.F. Homer-Dixon, J.H. Boutwell and G. Rathjens, 'Environmental Change and Violent Conflict', *Scientific American*, **268** (2), (1993), 16–23.
2. On this point see Urs Luterbacher and Ellen Wiegandt, *Modeling the Impact of Climate on Society*, CIESIN working paper, Ann Arbor, Mich., 1990.
3. There are numerous economic models of individual decisions. Slobodan Djajic is one of the economists who has studied the migration question most thoroughly. See, for instance, his 'Illegal Aliens, Unemployment and Immigration Policy', *Journal of Developmental Economics*, **25** (1987), 235–49.
4. This tension between individual and collective or governmental decisions regarding migration is emphasized in A. Zolberg, 'International Migrations in Political Perspective', in Mary Kritz, Charles Keely and Silvano Tomasi (eds), *Global Trends in Migration: Theory and Research on International Population Movements*, New York: Center for Migration Studies, 1981, pp. 3–27.

5. The vexed question of the aggregation of preferences is actually simpler to solve in most practical cases than people think. Lars-Erik Cederman shows that if a set of decision-makers have Prisoner's Dilemma, Chicken, Deadlock or Stag Hunt type preferences, majority rule will almost always lead to a consistent social preference order of the above types (cf. Lars-Erik Cederman, 'Unpacking the National Interest: an Analysis of Preference Aggregation in Ordinal Games', in this volume. This is particularly relevant in our case here, since, as we shall show, our analysis reveals *ordinal* Prisoner's Dilemma-type preferences for both countries' decision-makers.
6. In another model, Jean-Paul Theler deals with the question of a segmented labour market where migrations could go both ways: as skilled labour (which can then be considered a form of capital) from more-developed to less-develcped countries; as unskilled labour from developing to developed countries. His work appears in *Analyses et prévisions, 1990–1992,* Lausanne: Institut Créa de macroéconomie appliquée, Ch. II.
7. This is a standard result of macroeconomic analysis. For a demonstration, see, for instance, Thomas J. Sargent, *Macroeconomic Theory*, New York: Academic Press, 1986, p. 21.
8. For all this refer to Donald A. Pierre, *Optimization Theory with Applications,* New York: Dover, 1986.
9. The bracketed expression in the following equation is equivalent to $\partial R_N/\partial L_{SN}$. *Is* is obviously positive given our assumptions about negative second partial derivatives of the production function. This proves our assertions about $\partial Y_N/\partial L_{SN}$ and $\partial R_N/\partial L_{SN}$ in the beginning of the discussion on the static situation.

Index

Printed and bound by CPI Group (UK) Ltd, Croydon, CR0 4YY

16/04/2025

14658436-0001